Cooperative Marketing of Fruits and Vegetables in India

K.V. Subrahmanyam
Principal Scientist (Agricultrual Economics)
Indian Institute of Horticultural Research (ICAR),
Bangalore-560089

T.M. Gajanana
Scientist (SS) (Agricultrual Economics)
Indian Institute of Horticultural Research (ICAR),
Bangalore-560089

CONCEPT PUBLISHING COMPANY, NEW DELHI-110059

ISBN : 81-7022-821-2 (P.B.)

First Published 2000

Published and Printed by
Ashok Kumar Mittal
Concept Publishing Company
A/15-16, Commercial Block, Mohan Garden,
New Delhi-110059 (INDIA)

Phone : 5648039, 5649024
Fax : 091-11-5648053
Email : publishing@conceptpub.com

FOREWORD

Cooperative movement in India has a long history. Started with the simple objective of giving loans to farming community way back in 1904, the movement has undergone changes over time. Efforts were made from Second Five Years Plan onwards to promote cooperatives as an instruments of linking credit with marketing. The movement now encompasses various economic activities like marketing, processing, supply of production requisites etc. However, the movement could not make a dent and the record of cooperatives has been far from satisfactory as there have been more failures than successes in the performance of such organisations. This has been especially true in respect of cooperatives engaged in marketing of fruits and vegetables.

The cooperatives should play an important role as agents of decentralised production and centralised marketing and offer themselves as an effective alternate channel in marketing of fruits and vegetables because of the large scale exploitation by the middlemen due to their perishable and seasonal nature. Being an economist working in this field for the past several years, I always felt that there was a need to address these problems and also to ascertain the reasons for success and causes of failures of the cooperatives in marketing of fruits and vegetables. However, I rarely came across in-depth studies in this important area of research.

The study on *Cooperative Marketing of Fruits and Vegetables in India* by Dr. K.V. Subrahmanyam and Dr. T.M. Gajanana is a good attempt in this direction and the Book has come at an appropriate time in view of the emphasis laid on horticulture sector during Eighth and Ninth Five Year Plans. The case studies selected from different geographical locations in horticulturally important states like Karnataka, Maharashtra and Andhra Pradesh highlight both successful and not so successful efforts of cooperatives in marketing of fruits and vegetables. The factors responsible for the existence of this type of dichotomy have been studied in detail by the authors. The economics of value addition through small scale processing of fruits and vegetables in one of the case studies brought out by the authors shows the need and importance of this activity for the success of the cooperative marketing societies. The guidelines/suggestions for starting new or successfully running the existing cooperative marketing societies made by the authors based on in-depth analysis of the case studies will be very valuable for the progress of the cooperative movement.

I am sure the Book would be useful to graduate and undergraduate students as a text and will serve as a practical guide to policy-makers, planners and development departments who are engaged in running the cooperative organisations for marketing of fruits and vegetables.

S. BISALIAH
Vice-Chancellor
University of Agricultural Sciences
G.K.V.K., Bangalore-65

PREFACE

Due to the emphasis given during the Eighth Five Year Plan by Government of India on the development of horticultural crops by allocating around Rs. 1000 crores, the production has increased tremendously and India has occupied the second position in the world production of fruits and vegetables. This increased production should result in increased returns to the cultivators for sustaining the tempo and one of the major bottlenecks is exploitation by middlemen in marketing the produce. To avoid this exploitation by the middlemen the Government of India, through organisations like NCDC, NHB, APEDA etc. is encouraging the formation of fruit and vegetable marketing cooperative societies by giving financial support in the form of share capital contribution, subsidies, loans etc. As the cooperative movement has more failures than successes, in order to make it viable and popular in different states one needs to study them in detail and come out with guidelines to make them successful. The present book is an outcome of the efforts in this direction and is based on the project work undertaken by us on this aspect at Indian Institute of Horticultural Research (IIHR), Bangalore. The book is based on the in-depth analysis of detailed primary and secondary data of the fruit and vegetable marketing cooperatives.

The book is divided into 5 parts comprising 14 chapters. Part I consists of two chapters. Chapter one introduces the subject matter of the book by presenting status of production of fruits and vegetables in India in comparison with the world, plan outlays for horticulture and the emphasis given to horticulture during the Eighth Five Year Plan and its implications on production and marketing of fruits and vegetables etc. In chapter 2 the very objective of the Book i.e. need for studying cooperative marketing of fruits and vegetables is presented highlighting the favourable role played by cooperatives in reducing the number of middlemen thereby increasing the producer's share in consumer rupee.

Part II consists of chapters 3 and 4 which trace the origin and progress of cooperative movement in India since 1904 and present status of cooperative marketing of fruits and vegetables. The statewise analysis of value of fruits and vegetables handled, infrastructure facilities like cold storage, processing units available/required, role of cooperatives in export of fruits and vegetables and also in market intervention and price support operations etc. are also discussed in this part.

In Part III three main organisations viz. NCDC, NHB and APEDA which exclusively support cooperatives are studied in detail in chapters 5, 6 and 7 respectively. Aspects like their objectives and progress, various schemes designed by them for encouraging the domestic and export marketing of fruits and vegetables like share capital contribution by NCDC, soft loan facility by NHB, air freight subsidy for export by APEDA are discussed.

Part IV consists of 6 chapters (chapters 8 to 13) covering the need for in-depth studies of fruit and vegetables marketing cooperatives, methodology adopted for studying them and four case studies namely HOPCOMS, Bangalore, and DKHOPGCOMS, Mangalore from Karnataka, FRESH, Hyderabad from Andhra Pradesh and Cooperative Fruit Sale Societies, Jalgaon from Maharashtra.

Part V, consisting of chapter 14, gives some guidelines/suggestions for starting a new cooperative or for successfully running the existing Fruit and Vegetable marketing cooperatives based on the results of the analysis of the case studies.

Besides, the Book contains about 100 references and also subject index to help the readers to locate specific pages and topics.

We are extremely grateful to Dr. S. Bisaliah, Vice-Chancellor, University of Agricultural Sciences, Bangalore for writing the foreword to this book.

We wish to thank the Director, IIHR, Bangalore for providing the necessary facilities to carry out the work. We benefited from our discussion with Mrs. M. Sudha, Scientist (AgEcon), IIHR, Bangalore and are grateful to her for sharing her experiences on cooperatives especially on FRESH, Hyderabad. We are thankful to Mr. V. Dakshina Moorthy, Technical Officer, IIHR, Bangalore for his assistance in the preparation of the manuscripts. We are grateful to the chairmen, managing directors, managers and the staff of HOPCOMS, Bangalore; DKHOPGCOMS, Mangalore; FRESH, Hyderabad and Cooperative Fruit Sale Societies in Jalgaon district for their willing cooperation and help during data collection.

We hope the book will be useful both to the teaching institutions engaged in offering marketing courses as a text and as a practical guide to policy-makers and Development Departments of Horticulture and Agriculture who are engaged in starting and running the fruit and vegetable marketing cooperatives to help the farmers to enable them to get better prices for their produce.

K.V. SUBRAHMANYAM
T.M. GAJANANA

CONTENTS

Page

LIST OF TABLES

LIST OF FIGURES

PART I

INTRODUCTION

1

FRUIT AND VEGETABLE PRODUCTION IN INDIA AND ITS IMPLICATIONS

The importance of Horticulture in the field of agriculture has been realised of late only by the Indian planners. Horticulture is looked upon as the potential foreign exchange earner. This can be seen from the increased outlay of Rs. 10,000 million in Eighth Plan as compared to Rs. 76.1 million outlay in the Fifth Plan (Fig. 1.1). Besides the existing programmes in Eighth Plan a number of new programmes like floriculture, tuber crops, mushrooms, medicinal and aromatic crops, were introduced and a special provision of Rs. 975 million for exports has also been made in Eighth Plan. (Table 1.1). At micro level also there is a perceptible change and more and more cultivators have started growing horticultural crops on par with other commercial crops. Because of these, India's position in the world has undergone a remarkable change.

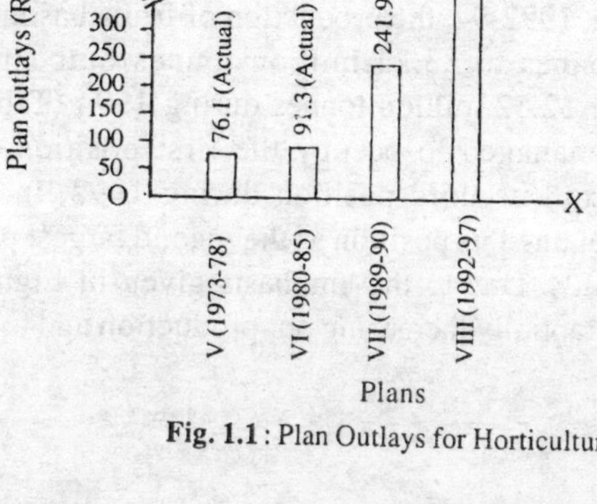

Fig. 1.1 : Plan Outlays for Horticulture

Table 1.1 : Plan Allocations for Horticulture during Seventh and Eighth Plans

(Rs. Million)

Items	*VII Plan*	*VIII Plan*
A. Ongoing		
Fruits	38.8	850
Vegetables	17.7	150
Coconut	77.1	1000
Spices	20.1	1500
Cashew	29.0	650
Cocoa	1.4	30
NHB	48.0	2000
Sub Total	232.1	6180
		(2563% increase)
B. New programmes		
Floriculture		100
Tuber crops		25
Mushrooms		100
Arecanut		50
Medicinal and Aromatic plants		50
Betelvine		20
Plastic		2500
Exports		975
Sub Total		3820
Grand Total	232.1	10000

Source : Kaul, 1993.

India's Position in World Production

India, though remained the second largest producer of fruits and vegetables (Figs. 1.2 and 1.3) has surpassed Brazil in the production of fruits, which used to occupy the first position during 1992. Within a period of three years i.e. 1992-94, the production of fruits has increased from 30.04 million tonnes to 33.24 million tonnes outbeating the Brazilian production of 32.52 million tonnes during 1994 (Table 1.2). Of course China has managed to occupy the first position in fruit production by 1994 from its third position during 1992. In case of vegetables, India has retained its position as the second largest producer next to China (Table 1.3). Due to the emphasis given in Eighth Plan there is a scope for substantially increasing our production and to surpass even China.

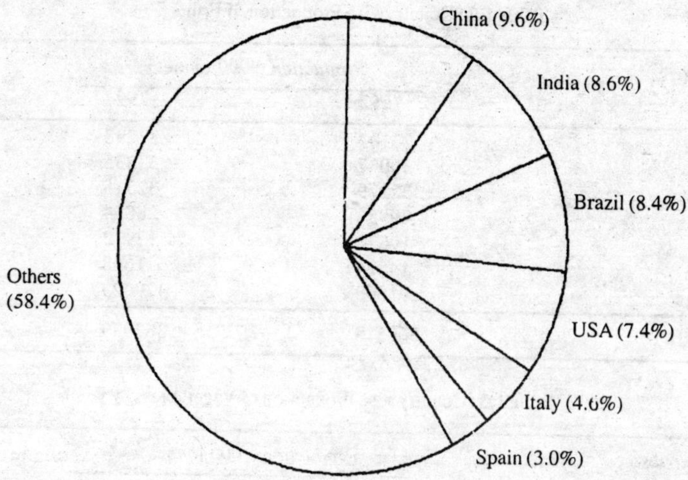

Fig. 1.2 : World Fruit Production

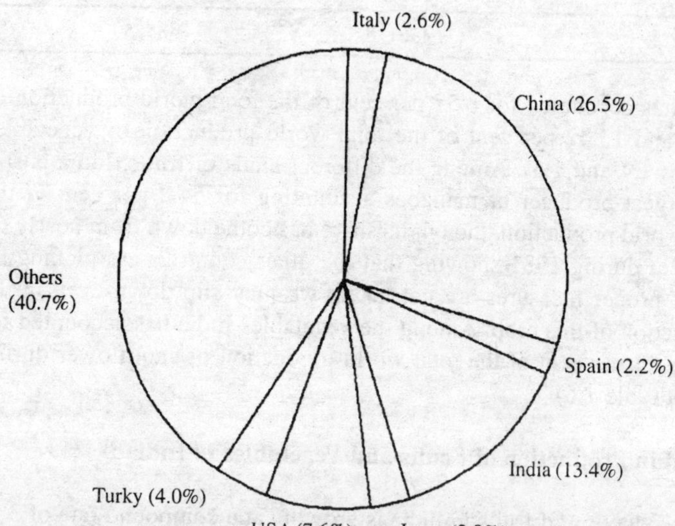

Fig. 1.3: World Vegetable Production

Table 1.2: Countrywise Production of Fruits

Countries	*Production ('000 tonnes)*	
	1992	*1994*
China	23102	37298
India	30037	33235
Brazil	32279	32515
USA	29956	28854
Italy	19820	17972
Spain	14524	11648
Others	291800	429023
World	369518	387939

Table 1.3: Countrywise Production of Vegetables

Countries	*Production ('000 tonnes)*	
	1992	*1994*
China	119786	128811
India	59194	65137
USA	30438	36443
Japan	13737	13870
Italy	14120	13629
Others	189735	413027
World	456170	485550

India has accounted for 8.57 per cent of the total world production of fruits and 13.42 per cent of the total world production of vegetables (Tables 1.4 and 1.5). Among the different kinds of fruits, India is still the largest producer of mangoes accounting for 54.2 per cent of the total world production, though its share has come down from nearly 65 per cent during 1981 showing that the other countries are picking up and if proper measures are not taken, we may slip down even in the production of this crop. Among the vegetables India has accounted for nearly 44 per cent of the total world production of cauliflower during 1994 (Table 1.5).

Trend in Production of Fruits and Vegetables in India

The production of fruits in India is growing at a compound rate of 3.82 per cent during the last one decade i.e. 1984-85 to 1994-95 and if this trend continues we may be able to produce 41.62 million tonnes of fruits by the end of the century i.e. 2000 A.D. (Table 1.6). Even this is short of

Table 1.4: India's Share in the World Production of Fruits

Fruits	*Production ('000 tonnes)*		*India's Share (%)*
	India	*World*	
Mango	10000	18450	54.20
Banana	7900	58584	15.02
Oranges	2100	58731	3.52
Apple	1238	48890	2.53
Papaya	1500	6200	24.19
Pineapple	820	11832	6.93
Grapes	750	56392	1.33
Pears	140	11231	1.25
Peaches	84	10935	0.80
Plums	58	7261	0.80
All fruits	33235	387939	8.57

Source : FAO Production Year Book, 1994.

Table 1.5: India's Share in the World Production of Vegetables

Vegetables	*Production ('000 tonnes)*		*India's Share (%)*
	India	*World*	
Potatoes	15000	265436	5.65
Tomatoes	5029	77540	6.49
Cauliflower	4800	10888	44.08
Onions	4300	32546	13.21
Cabbage	3300	40250	8.20
Green peas	268	4346	6.17
All vegetables	65137	485550	13.42

Source : FAO Production Year Book, 1994.

the target fixed by the National Commission on Agriculture at 45 million tonnes to meet the dietary requirements (Fig. 1.4).

Regarding the growth of individual fruit crops, varying growth patterns can be noticed (Table 1.7). During the last decade (1982-83 to 1992-93) the maximum positive growth of area of 182 per cent was noticed in case of grapes followed by citrus fruits. In contrast two important fruits guava and pineapple have recorded negative growth rates in area. On the whole the area under fruits has registered a positive growth of 25.61 per cent in the last decade. Regarding production of fruits, the increase was nearly 45 per cent, with grapes, banana and citrus fruits being the main contributors.

Table 1.6: Fruit and Vegetable Production in India

(*Million tonnes*)

Year	*Fruits*	*Vegetables*	*Total*
1984-85	23.26	35.84	59.1
1985-86	24.62	47.32	71.94
1986-87	24.21	48.51	72.72
1987-88	26.5	48.1	74.6
1988-89	27.38	49.26	76.64
1989-90	28.00	53.89	79.89
1990-91	28.4	53.01	81.41
1991-92	32.07	66.58	98.65
1992-93	32.96	71.01	103.97
1993-94	32.45	63.80	96.25
1994-95	33.24	65.14	98.38
CGR (%)	3.82	5.65	5.06
2000-01*	41.62	90.59	132.21
2010-11*	48.36	112.86	161.22
2020-21*	52.12	125.97	178.09

*Projection based on CGR with 1994-95 as base year.

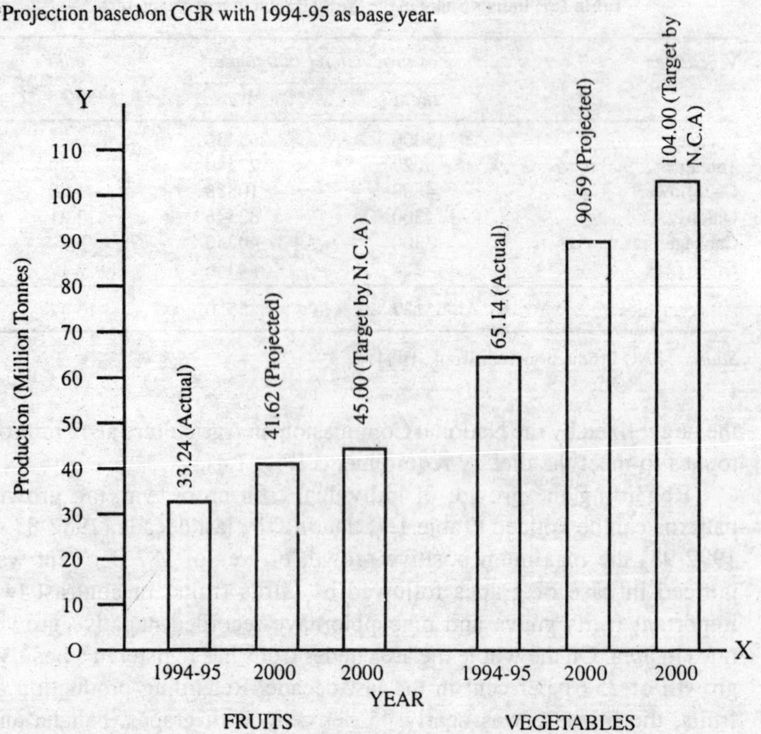

Fig. 1.4: Projected Fruit and Vegetable Production and Requirement by 2000 A.D.

Table 1.7 : Changes in Area and Production of Different Fruits

Fruits	Area ('000 ha)					Production ('000 tonnes)				
	1982-83		1992-93		Percentage change over 1982-83	1982-83		1992-93		Percentage change over 1982-83
	Area	%	Area	%		Production	%	Production	%	
Banana	317.6	12.45	396.2	12.36	24.75	5608.03	24.75	10459.99	31.74	86.52
Mango	1022.22	40.05	1136.67	35.46	11.2	8833.80	38.98	9223.26	27.99	4.41
Citrus	234.57	9.19	367.56	11.53	57.55	1807.89	7.98	2979.11	9.04	64.78
Guava	155.78	6.10	111.85	3.49	-28.2	1350.40	5.96	1204.31	3.65	-10.82
Apple	161.71	6.34	191.26	5.97	18.27	966.60	4.26	1168.25	3.54	20.86
Pineapple	87.2	3.42	59.44	1.84	-31.83	671.82	2.96	858.98	2.61	27.86
Papaya	NA		47.43	1.48				803.60	2.44	
Grapes	12.05	0.47	34.01	1.06	182.24	237.02	1.05	653.16	1.98	175.57
Sapota	NA		30.78	0.96				423.14	1.28	
Litchi	NA		53.56	1.67				260.94	0.79	
Others	560.88	21.98	774.91	24.18	38.16	3185.87	14.06	4920.53	14.94	54.45
All India	2552.01	100	3205.65	100	25.61	22661.43	100	32955.28	100	45.42

Source : 1982-83—Subrahmanyam, 1985.
1992-93—NHB, 1995.
NA = Not available.

In case of vegetables, during the last 11 years period, i.e. 1981-82 to 1992-93, the variation in area was marginal. In 1992-93 while the area has decreased as compared to 1981-82, during 1991-92 the area has increased as compared to 1981-82. These fluctuations in areas are expected as vegetables are field crops occupying the land for a period of nearly 3 to 4 months. There was a definite positive increase in production to the extent of nearly 34 per cent in 1992-93 as compared to 1981-82 (Table 1.8).

On the whole there is a positive rising trend in production of both vegetables and fruits in the country during the last ten years (Fig. 1.5).

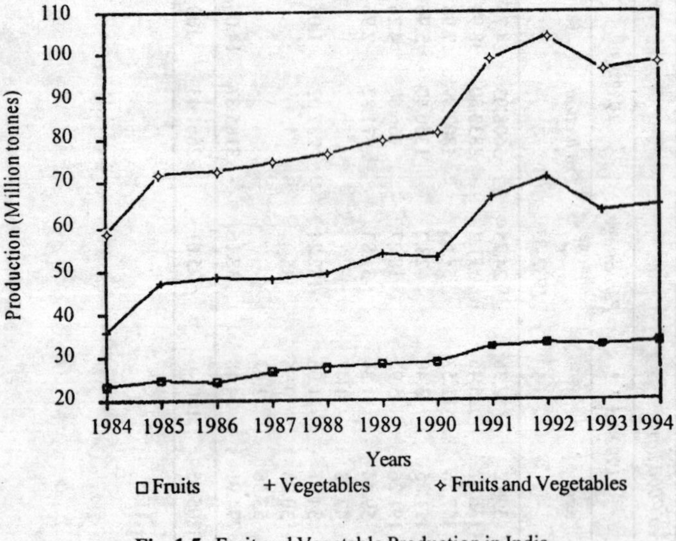

Fig. 1.5 : Fruit and Vegetable Production in India

Contribution of Different States

The contribution of different states of India towards the area and production of fruits and vegetables is in Tables 1.9 and 1.10. The states of Andhra Pradesh, Maharashtra, Karnataka, Bihar and U.P. are the major contributors towards area and production of fruits during 1992-93 whereas Bihar, West Bengal, U.P. and Orissa are the major contributors in the production of vegetables.

Table 1.8 : Changes in Area and Production of Vegetables in India

Vegetables	Area ('000 ha)							Production ('000 tonnes)						
	1981-82		1991-92		1992-93		% change over 1981-82	1981-82		1991-92		1992-93		% change over 1981-82
	Area	%	Area	%	Area	%		Pro-duction	%	Pro-duction	%	Pro-duction	%	
Potato	NA	NA	289.08	5.17	308.65	6.05		NA	NA	4244.37	7.24	4849.57	6.83	
Cabbage	NA	NA	178.35	3.19	215.86	4.23		NA	NA	2796.43	4.77	4357.05	6.14	
Cauliflower	NA	NA	202.79	3.62	222.44	4.36		NA	NA	2998.06	5.12	4220.85	5.94	
Okra	NA	NA	221.99	3.96	281.63	5.52		NA	NA	1887.08	3.22	2486.98	3.50	
Peas	NA	NA	177.68	3.17	194.15	3.80		NA	NA	1296.01	2.21	1491.97	2.10	
Others	591.00*	10.83	3063.28	54.70	2239.16	43.89		7446.30*	14.02	22471.66	38.36	27596.86	38.87	
Total fresh Vegetables	4443.00	81.43	4133.17	73.81	3461.89	67.85	-22.08	40554.30	76.35	35693.61	60.92	45003.27	63.38	10
Potato	762.70	13.98	1135.08	20.27	1259.92	24.69	65.19	9911.80	18.66	18192.98	31.05	20284.37	28.57	104
Onion	250.70	4.59	331.74	5.92	380.87	7.46	51.92	2647.90	4.99	4705.66	8.03	5715.41	8.05	115
All India	5456.40	100.00	5599.99	100.00	5102.68	100.00	-6.49	53114.00	100.00	58592.25	100.00	71006.05	100.00	33

Source : 1981-82—Singh, 1983.
1991-92 and 1992-93—NHB, 1995.

NA = Not available

* Other tuber and bulb crops.

Table 1.9 : Share of States in Area and Production of Fruits (1992-93)

Area : hectares
Production : tonnes

States	*Area*	*%*	*Production*	*%*
A.P.	338512	10.56	4766421	14.46
Maharashtra	294014	9.17	4313830	13.09
Karnataka	225389	7.03	3191835	10.43
Bihar	271170	8.46	2799250	10.19
U.P.	481290	15.02	2278487	10.18
Tamil Nadu	136908	4.27	2316100	7.19
Kerala	342000	10.67	1101296	5.86
Gujarat	93500	2.92	1828880	5.81
M.P.	68403	2.13	1245000	3.89
W. Bengal	133890	4.18	1131680	3.49
Assam	98323	3.07	886381	3.35
Orissa	149452	4.66	978000	3.14
J & K	184621	5.76	700802	2.39
Punjab	76519	2.39	663808	2.10
Others	311658	9.72	1457983	4.42
Total	3205649	100.00	32955276	100.00

Source : NHB, 1995. Horticultural Data Base, 1992-93.

Table 1.10 : Share of States in Area and Production of Vegetables (1992-93)

Area : hectares
Production : tonnes

States	*Area*	*%*	*Production*	*%*
Bihar	887309	17.39	16200449	22.82
W. Bengal	609990	11.95	12976880	18.28
U.P.	672915	13.19	9598199	13.52
Orissa	760660	14.91	7745000	10.91
Karnataka	416958	8.17	4662172	6.57
Maharashtra	201633	3.95	3570887	5.03
Kerala	202503	3.97	2902922	4.09
M.P.	214049	4.19	2046852	2.88
Assam	191900	3.76	1754355	2.47
Gujarat	142000	2.79	1556500	2.19
Punjab	85299	1.67	1455300	2.05
A.P.	148519	2.91	1368068	1.93
Haryana	73581	1.44	1029530	1.45
Tamil Nadu	53530	1.05	959073	1.35
J & K	180300	3.53	745000	1.05
Delhi	47217	0.92	7357664	1.03
Others	214312	4.20	1699701	2.39
Total	5102675	100.00	71006052	100.00

Source : NHB, 1995, Horticultural Data Base, 1992-93.

For mango, Andhra Pradesh, Uttar Pradesh, Bihar and Karnataka are the important production centres whereas Bihar and Maharashtra are important for citrus fruits. Maharashtra is the most important centre of production for grapes followed by Karnataka (Table 1.11). In case of vegetables, U.P., West Bengal and Bihar are the top producers of potato whereas for onion, Maharashtra and Karnataka were the major producers. For other seasonal vegetables, different states occupied different positions (Table 1.12).

Per Capita Availability

Inspite of the above tremendous trends in the production of fruits and vegetables in the country, gross per capita availability is far less than the requirements based on ICMR recommendations (Table 1.13). Besides the low production and productivity the huge post-harvest losses due to inefficient marketing is also responsible for the low per capita availability. Hence there is a need to improve production not only by technological developments but also by increasing the efficiency of marketing which is discussed below.

Need for Efficient Marketing

The existing marketing system has not matched the tremendous growth in production of fruits and vegetables, which has been evident from the huge post-harvest losses of fruits and vegetables while marketing them (Table 1.14). Besides, for sustaining the interest of the cultivators and motivating them for higher production it is necessary to ensure that the cultivators get a reasonable price as many studies on marketing of fruits and vegetables have shown that the share of the producer in consumer rupee is very low due to too many marketing intermediaries and concentration of trade in a few hands. Hence it is necessary to identify a marketing channel which assures a good and reasonable price to the producers as well as consumers with least number of intermediaries. It is in this context that the co-operative marketing comes into play and the present study aims at focusing on this aspect.

Table 1.11 : Share of States in the Production of Different Fruits

(*Percentages*)

States	*Mango*	*Banana*	*Citrus*	*Guava*	*Apple*	*Pine-apple*	*Papaya*	*Grapes*	*Sapota*	*Litchi*	*All fruits*
A.P.	29.29	9.49	28.28	6.19	—	—	0.37	8.63	11.42	—	14.46
Maharashtra	3.55	26.05	14.06	4.10	—	—	4.58	51.54	6.14	—	13.09
Karnataka	7.57	11.25	10.70	11.05	—	12.53	22.36	19.67	63.10	0.02	10.43
Bihar	19.13	4.60	5.81	20.88	—	7.88	13.07	—	0.11	76.64	10.19
U.P.	20.34	0.20	2.02	14.23	17.98	—	1.96	0.12	—	4.36	10.18
Tamil Nadu	3.79	16.00	5.72	6.94	—	3.35	—	7.17	—	—	7.19
Kerala	2.67	3.76	—	—	—	5.47	6.97	—	—	—	5.86
Gujarat	3.69	10.71	4.30	8.22	—	—	—	—	—	—	5.81
M.P.	2.12	6.76	5.13	12.05	—	—	4.67	—	—	—	3.89
W. Bengal	1.75	1.72	0.81	3.86	—	25.90	18.35	—	—	6.45	3.49
Assam	0.05	5.15	2.86	3.16	—	21.48	10.34	—	0.35	4.60	3.35
Orissa	3.25	2.79	2.01	2.41	—	—	15.55	—	1.91	1.52	3.14
J & K	—	—	—	0.18	57.29	—	0.02	—	—	—	2.39
Punjab	0.87	—	10.27	3.90	—	—	—	9.32	—	3.85	2.10
H.P.	0.17	—	0.31	0.06	23.89	—	0.07	—	—	—	—
Arunachal	—	—	—	—	1.59	—	0.12	—	—	—	—
Manipur	—	—	—	—	7.04	—	—	—	—	—	—
Meghalaya	—	—	—	—	8.44	—	—	—	—	—	—
Tripura	—	—	—	—	3.75	—	—	—	—	—	—

Source : NHB, 1995.

Table 1.12 : Share of States in the Production of Different Vegetables

(Percentages)

States	*Potato*	*Onion*	*Tomato*	*Cabbage*	*Cauliflower*	*Okra*	*Peas*	*All vegetables*
Bihar	26.82	16.02	15.70	15.43	24.31	29.29	2.15	22.82
W. Bengal	24.37	—	9.28	25.25	26.06	16.08	1.61	18.28
U.P.	30.03	7.88	1.39	11.92	17.96	0.45	61.75	13.52
Orissa	0.59	5.86	9.79	14.46	12.94	27.18	3.42	10.91
Karnataka	3.18	18.98	12.88	5.33	—	3.49	1.08	6.57
Kerala	0.02	—	—	—	—	0.58	—	5.03
Maharashtra	0.97	27.59	11.41	4.85	2.88	2.46	—	4.09
M.P.	1.90	3.72	6.91	2.75	6.70	2.60	15.42	2.88
Assam	1.91	0.22	5.00	8.59	4.48	4.75	1.03	2.47
Gujarat	1.91	8.45	1.86	2.52	2.49	1.17	—	2.19
Punjab	4.06	0.62	2.28	—	1.21	0.25	—	2.05
A.P.	0.04	2.87	8.31	0.26	—	6.59	0.19	1.93
Haryana	1.17	1.18	1.94	3.13	—	1.75	4.75	1.45
Tamil Nadu	0.63	3.04	9.18	1.44	0.13	1.17	0.01	1.35
J & K	—	18.98	—	—	—	—	—	1.05
Delhi	0.17	0.91	1.76	1.96	—	1.28	2.31	1.03
Arunachal	6.55	—	0.01	3.95	0.08	3.44	0.14	—
H.P.	—	—	—	—	—	—	4.44	—

Source : NHB, 1995.

Table 1.13 : Per Capita Availability and Requirements of Fruits and Vegetables

(Grams/person/day)

Year	*Gross Per Capital Availability*		
	Fruits	*Vegetables*	*Total*
Availability			
1975	74	154	228
1980	74	162	236
1985	89	173	262
1990	92	169	259
1994*	99	194	293
Requirement**	120	300	420

*Average of 1993 and 1994.

**Based on ICMR recommendation.

Table 1.14 : Post-harvest Losses in Major Fruits and Vegetables

Crop	*Post-harvest loss (%)*
Mango	17.1 - 36.7
Banana	11.0 - 14.0
Oranges	8.3 - 30.7
All fruits	6.0 - 36.7
Onion	13.0 - 30.0
Tomato	13.0 - 16.0
All vegetables	13.0 - 30.0

Source : IIHR, 1992.

2

NEED FOR COOPERATIVE MARKETING

In case of perishable commodities for a producer, the real problem starts when he attempts to dispose of the same i.e. while marketing the produce. Very few people realise that the cultivators have to spend a considerable amount for marketing the fruits and vegetables. Besides the huge expenditure the marketing cost also affects the income of the cultivators (*Subrahmanyam, 1981a*).

The marketing of fruits and vegetables, unlike in case of cereals and pulses, is more complex because of the special characteristics like highly perishable nature, seasonality, bulkiness etc. and needs special care and immediate disposal. The studies conducted on marketing of fruits and vegetables have shown the exploitation by the middlemen resulting in very low share of consumers rupee (Ayyasamy, 1980, Bhale Rao *et al.*, 1981, Patil *et al.*, 1985). Hence it is essential to critically analyse the present marketing system and identify the most suitable channel of marketing which will be beneficial to both the producers and consumers, the present weakness of this channel and explore the ways and means to strengthen the same.

Present Marketing System of Fruits and Vegetables

A. Fruits

India produces a variety of fruits ranging from temperate apples to tropical *ber*. But most of the production of these fruits are highly seasonal and their availability is confined to a few months in a year. For example the peak period of availability of mango, the king of fruits, is hardly confined to a two-month period i.e. May-June, oranges during November-January, papaya during February-March etc. (Fig. 2.1). This short period of availability in large quantities makes the marketing a difficult task with problems like storage, wide price fluctuations etc. Besides, the huge post-harvest losses because of perishable nature results not only in less per capita availability but also increases the cost of transport etc. (*Subrahmanyam, 1986a*). All these factors lead to exploitation by middlemen resulting in poor returns to cultivators, though consumers pay high prices. To overcome these problems the present system of marketing is examined briefly below to identify the desirable channel.

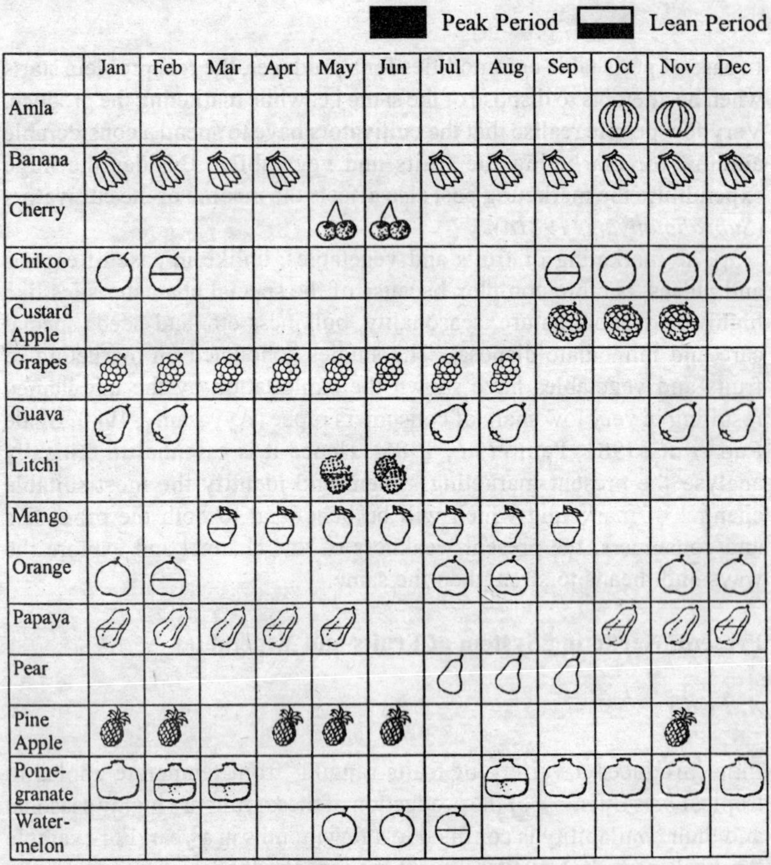

Fig. 2.1 : Fruit Seasons of India

Source : APEDA.

The marketing channels followed in case of fruits crops in India is presented in Fig. 2.2. The most popular channel used by the cultivators, given in the centre is:

> Producer—Pre-harvest contractor—Commission Agent—Wholesaler—Retailer—Consumer

The popularity of this channel was brought out in many studies on different fruits in different states over a period of time. The studies conducted on marketing of citrus fruits in 1973 in Punjab (*Mehta and Balwinder Singh, 1973*) has shown that nearly 84 per cent of the mandarin and 85 per cent of the orange fruits was marketed by the pre-harvest contractors. Similarly in case of mango fruits (*Balwinder Singh and Sidhu, 1976*) 73 per cent of the cultivators in Punjab have sold their crop to pre-harvest contractors. The studies conducted in three southern states viz. Karnataka, Andhra Pradesh and Tamil Nadu on important fruits in 1979-80 (*Subrahmanyam, 1983*) has once again shown the dominance of pre-harvest contractors in fruit trade, though the extent of sale has differed from state to state and fruit to fruit (Table 2.1). The recent studies conducted on marketing of mango in Andhra Pradesh (*Das and S. Ghosh, 1996*) have shown that nearly 80 per cent of the produce is sold to pre-harvest contractors. From this it is clear that over years, the dominance of pre-harvest contractors in fruit trade is still continuing, inspite of the fact that many studies have established that it is not a desirable practice (*Subrahmanyam, 1981b, 1986b*). The analysis of costs and returns have clearly shown (Table 2.2) that by taking up self marketing, the additional returns realised varied from 43 per cent to more than 100 per cent.

The next important marketing channel of marketing followed by the cultivators was selling through commission agents. The main problem in this marketing channel was cheating, illegal deductions towards charities, weighment etc. This is happening in case of regulated markets itself. The study conducted in four regulated markets viz. Delhi, Chandigarh, Agra and Jaipur, all charges payable by cultivators were found to be in excess of the prescribed rates (*Swarup and Singh 1987*). The produce sold through other channels of marketing was very negligible.

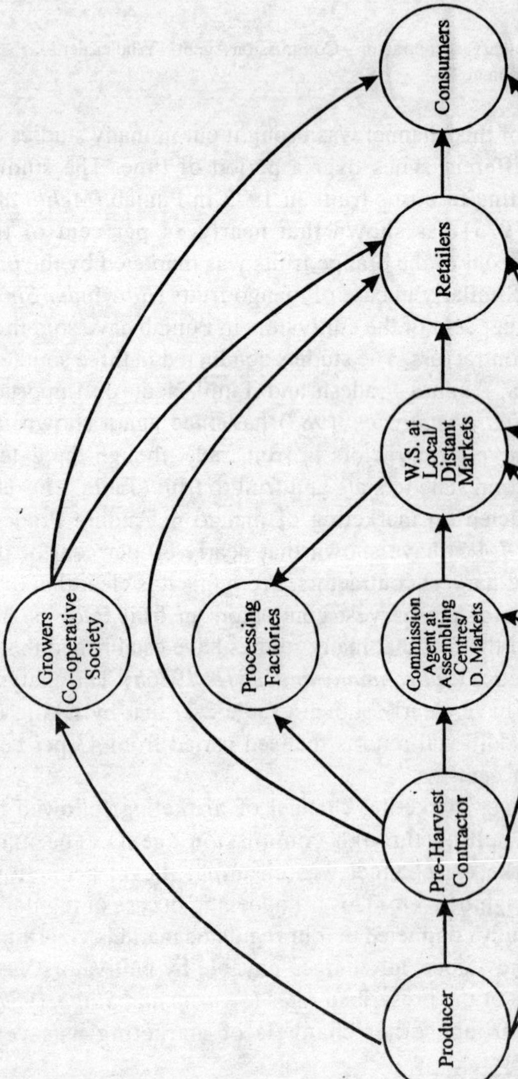

Fig. 2.2: Marketing Channels of Fruits in India

Table 2.1 : Distribution of Fruit Cultivators using Different Marketing Channels

(Percentage)

Merketing channel	*Mango*			*Banana*			*Citrus*				*Average*
	KNK	*AP*	*TN*	*KNK*	*AP*	*TN*	*KNK*	*AP*		*TN*	
							(Mandarin)	*Lime*	*Sweet orange*	*(Mandarin)*	
Producer-Pre-harvest contractor	100.00	64.10	56.41	61.43	79.89	93.75	84.75	72.22	94.44	61.11	73.60
Producer-Commission Agent (Market)	—	35.90	28.20	8.57	17.94	6.25	17.87	16.67	—	38.89	14.48
Producer-Whole Saler (Market)	—	—	5.13	28.57	—	—	1.69	11.11	5.56	—	10.22
Producer-Retailer (Field)	—	—	10.26	1.43	2.17	—	1.69	—	—	—	1.70

*KNK : Karnataka, AP : Andhra Pradesh, TN : Tamil Nadu.

Table 2.2 : Comparison of Costs and Returns from Self Marketing and Contract Sale in Mango in Different States

(Rs/Ha)

	Particulars	Andhra Pradesh		Tamil Nadu		Maharashtra	
		Self marketing	*Contract sale*	*Self marketing*	*Contract sale*	*Self marketing*	*Contract sale*
1.	Gross returns	7237.5	2162.25	7750.13	2131.72	15682	5850
2.	Extra cost/Savings in cost						
	(a) Maintenance	343.75	—	820.53	—	4832	1792
	(b) Marketing	1148.00	—	1548.43	—	5053	—
	Total	1491.75	—	2368.96	—	9885	1792
3.	Returns over item 2	5745.79	2162.25	5381.17	2131.72	5797	4058
4.	Percentage of additional returns	165.73	—	152.44	—	42.85	—

Source : For Andhra Pradesh and Tamil Nadu, Subrahmanyam, 1989 and for Maharashtra, Thorat *et al.*, 1990.

B. Vegetables

As in case of fruits, the vegetables are also highly seasonal and different vegetables come to market in different periods (Fig. 2.3). In case of vegetables, the adjustment of production i.e. area planted, is much easier to price changes, as they are field crops with a growing period of hardly 3 to 4 months. Because of this seasonal production will be very high resulting in wide price fluctuations. For example, the tomato prices, during 1995-96 in Bangalore market varied from as low as Rs. 217 per quintal in April to as high as Rs. 1021per quintal in July, nearly a five fold increase, within a span of three months (Fig. 2.4). Adding to this

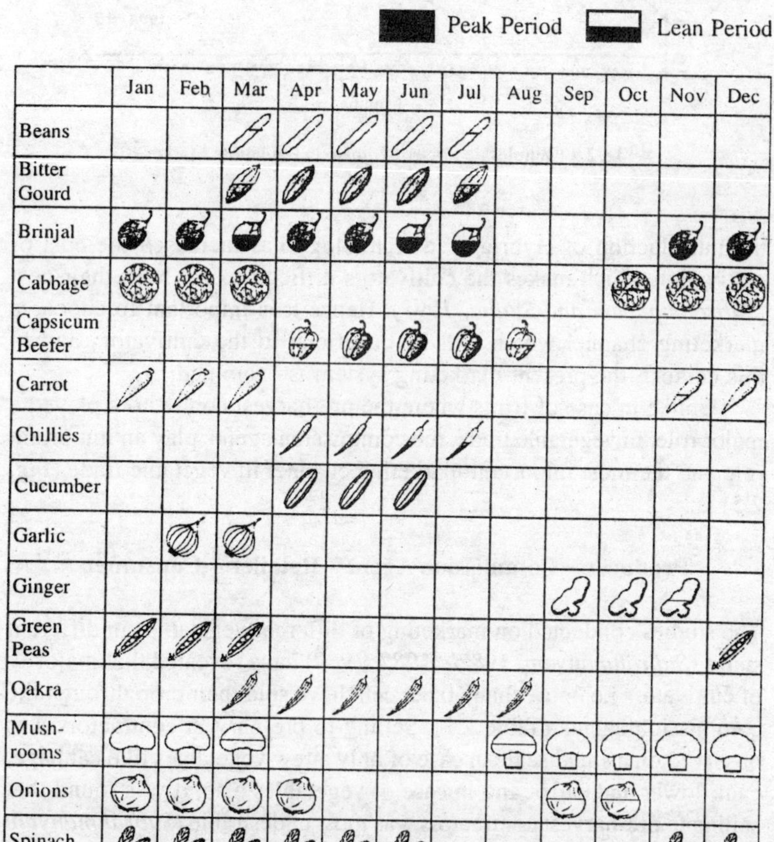

Fig. 2.3 : Vegetable Seasons of India

Source : APEDA.

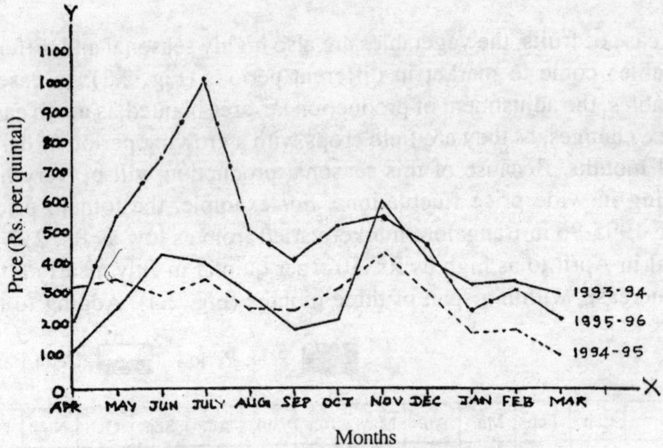

Fig. 2.4 : Wholesale Price of Tomato in Bangalore Market

the introduction of Hybrid seed technology has increased the cost of cultivation which makes the cultivators difficult to get back their cost (*Subrahmanyam and Sudha, 1993*). Hence it is important to choose a marketing channel which will be beneficial to the cultivators and in this context, the present marketing system is examined.

Unlike in case of fruits where the pre-harvest contractors played a major role, in vegetable trade, the commission agents play an important role and the most important marketing channel in vegetable trade (Fig. 2.5) is:

Producer—Commission Agent—Retailer—Consumer

The studies conducted on marketing of different vegetables in different states (*Subrahmanyam, 1982b, 1987, 1989*) have revealed that majority of cultivators i.e. more than 60 per cent have sold their crop through the commission agents (Table 2.3). Selling to pre-harvest contractors was observed on a small scale in case of only a few vegetables like cabbage, cauliflower and carrot and in case of vegetables also, it was found that selling to pre-harvest contractors was most undesirable (*Subrahmanyam 1982a*).

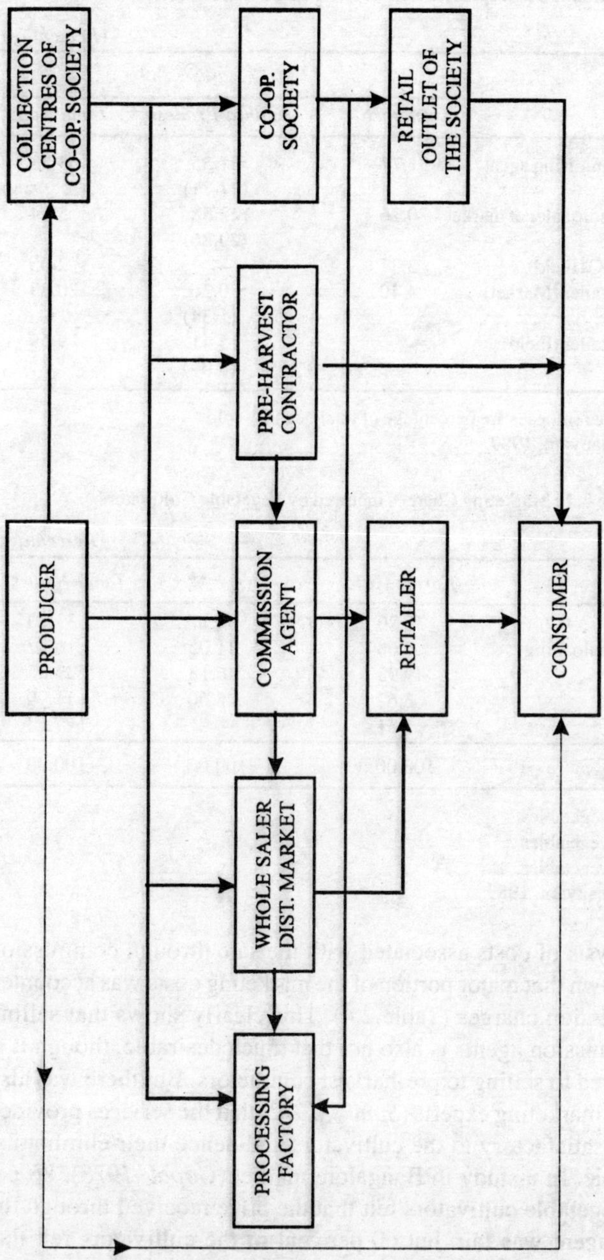

Fig. 2.5 : Vegetable Marketing Channels in India

Table 2.3 : Distribution of Vegetable Cultivators Using Different Marketing Channels

(Percentage)

Marketing Channel	*States*		
	Karnataka	*Andhra Pradesh*	*Tamil Nadu*
1. Producer—Commision agent	91.77	61.55 (77.71)	63.33
2. Producer—Whole saler at market	0.26	22.58 (20.86)	5.14
3. Producer—PHC (Field)	3.37	—	2.05
4. Producer—Retailer (Market)	4.40	0.36 (1.34)	20.89
5. Producer—Retailer (Field)	—	15.41 (0.08)	9.59

Note : Figures in parentheses are percentage of total quantity sold.
Source : Subrahmanyam, 1989.

Table 2.4 : Marketing Charges incurred by Vegetable Cultivators

(Percentage)

Particulars	*Karnataka**	*A.P.***	*Tamil Nadu****
1. Transport	38.26	34.17	33.90
2. Loading and unloading	6.68	11.05	6.87
3. Commission	49.93	46.18	39.65
4. Packing	3.62	8.60	11.90
5. Others	1.51	—	7.53
Total	100.00	100.00	100.00

*Average of 7 vegetables.
**Average of 4 vegetables.
***Average of 6 vegetables.
Source : Subrahmanyam, 1989.

The analysis of costs associated with the sale through commission agents has shown that major portion of the marketing costs was accounted by the commission charges (Table 2.4). This clearly shows that selling through commission agents is also not that much desirable, though it is better compared to selling to pre-harvest contractors. But there was also an opinion by marketing experts (*Spinks, 1972*) that the services provided by them was satisfactory to the cultivators and hence their elimination is not desirable. In a study in Bangalore market (*Gopal, 1978*), 86 per cent of the vegetable cultivators felt that the price received through the commission agent was fair, but 60 per cent of the cultivators felt that the commission charges are a bit high. Hence the measures taken should not be aimed at eliminating the commission agents/market intermediaries completely, but to control their activities, through laying down rules and regulations and also in offering an alternate channel of sale to cultivators, which is more beneficial and attractive. It is in this context that co-operative marketing has to be developed and encouraged to play a greater role.

Benefits of Cooperative Marketing

As an alternate channel of marketing to reduce the number of intermediaries and benefit both producers and consumers, the establishment of cooperative marketing societies is generally advocated. In fact, the All India Rural Credit Survey Committee, way back in 1954 itself, has said "the cooperation has failed but cooperation must succeed" This statement has assumed much more relevance and significance in the present context especially with respect to marketing of perishables like fruits and vegetables. In general where the Fruit and Vegetable Cooperative Marketing Societies are functioning well, the producers are benefited by marketing through these societies. It is evident from Table 2.5 that the cooperatives are found to be the best channel of marketing mango, next only to direct sale to consumers, as the growers realised the maximum share of 74 per cent in the consumer price as against only 40 per cent and 28 per cent when they chose to sell through commission agents and PHCs respectively. The consumers also benefited from cooperatives as they paid only Rs. 355 per crate in cooperative retail outlets against Rs. 455 per crate when they purchased from other agencies (Table 2.5). In Jalgaon, the banana growers realised Rs. 97 per quintal by selling through cooperatives but could get only Rs. 80 per quintal by selling through other agencies (Table 2.6). Similarly, the chikoo growers in Gujarat could get higher prices by selling through cooperative societies compared to other agencies (*Parmar et al., 1989*). It may be seen from Table 2.7 that in case of vegetables, the wholesale price realised by the brinjal growers in Karnataka was higher (Rs. 51.45 per quintal) when it was sold through cooperatives than when they marketed it through commission agents at Rs. 46.27 per quintal. The share of producers selling through the cooperatives was 61 per cent as against 55 per cent in case of those selling through the commission agents. The Horticultural Producers' Cooperative Marketing and Processing Society Ltd. (HOPCOMS) in Karnataka helped the producers by offering favourable wholesale prices compared to other agencies and

Table 2.5 : Producer's Share, Marketing Cost and Consumer's Price in different Channels of Marketing Mango

(*Rs. per crate*)

Channel	*Producer's share*	*Marketing cost*	*Consumer's price*
1. Producer—Consumer	161.27 (99.35)	0.73 (0.45)	162.00 (100)
2. Producer—Cooperative Society—Consumer	246.00 (74.37)	53.23 (14.99)	355.00 (100)
3. Producer—Commission Agent—Wholesaler Retailer—Consumer	182.00 (40.00)	75.78 (16.66)	455.00 (100)
4. Producer—PHC—Commission Agent—Wholesaler—Retailer—Consumer	128.00 (28.13)	75.78 (16.86)	455.00 (100)

Source : Wadkar *et al.*, 1994.

Table 2.6 : Price Realised by Banana Growers in Different Marketing Channels

Channels	*Price realised (Rs/q)*		
	Small	*Medium*	*Large*
1. Producer—Cooperative Fruit Sales Society	96.80	86.63	92.57
2. Producer—Group Sale Agency	79.14	91.09	84.85
3. Producer—Private traders	80.06	81.49	86.99

Source : Pawar, 1991.

Table 2.7 : Price Realised, Producer's Share and Coefficient of Variation in Wholesale Prices in Different Channels of Marketing of Brinjal

Channels	*Price realised (Rs/q)*	*Producer's share (%)*	*Coefficient of variation (%)*
1. Producer—Commission Agent—Retailer—Consumer	46.27	55.00	11.90
2. Producer—Cooperative Society—Retailer—Consumer	51.45	61.00	6.90

Source : Hugar *et al.*, 1983.

the consumers by charging a reasonable price in its retail outlets (*Narasimhan and Ravindra, 1989*). Besides, it may also be noted from Table 2.8 that in marketing of cabbage and carrots in Tamil Nadu, the producer's share was higher (51 per cent and 55 per cent for cabbage and 68 per cent and 72 per cent for carrots) in channels where cooperative organisations were involved (*Selvaraj and Krishnamoorthy, 1990*).

Table 2.8 : Producer's Share in Marketing of Vegetables

(in percentage)

	Marketing Channels	*Producer's share (%)*	
		Cabbage	*Carrot*
1.	Producer—NCMS—Wholesaler—Retailer—Consumer	51.44	68.09
2.	Producer—Mandies—Wholesaler—Retailer—Consumer	48.67	66.36
3.	Producer—NVGC—Wholesaler—Retailer—Consumer	55.42	72.01
4.	Producer—Pre-harvest contractor—Wholesaler—Retailer—Consumer	30.46	56.62

Source : Selvaraj and Krishnamoorthy, 1990.
Note : NCMS—Nilgiris Cooperative Marketing Society.
NVGC—Nilgiris Vegetable Growers' Cooperative.

From the above discussion it is clear that the cooperative marketing societies can be looked as the best alternative channel which can look after the interest of the producers. But, inspite of this the extent of use of this channel is very low. In majority of the places this channel does not even exist and in a few places where this channel is available the extent of its use is very limited. For example in case of mango marketing in Maharashtra hardly 0.83 per cent of the cultivators has used this channel and only 0.16 per cent of the produce was sold through this channel. (*Wadkar et al., 1994*). The non-popularity of the cooperative marketing societies was also evident from the poor and stagnating membership in the societies (*Subrahmanyam et al., 1994; Sudha and Subrahmanyam, 1993*).

Realising the importance of cooperatives in the field of marketing of fruits and vegetables, of late, organisations like National Cooperative Development Corporation (NCDC), National Horticultural Board (NHB), APEDA etc., are encouraging the formation of such cooperatives with liberal financial assistance in the form of contribution towards share capital, working capital, soft loan, subsidies etc. However, inspite of this the fruit and vegetable cooperatives have not become popular and as we shall see later, the performance of such cooperatives differed from state to state (perhaps due to the policies followed by them) resulting in the existence of both successful and not so successful fruit and vegetable cooperatives, the latter being more.

PART II

PRESENT STATUS

3

ORIGIN AND PROGRESS OF COOPERATIVE MOVEMENT IN INDIA

Though the cooperative movement in the world started with consumer cooperatives (*Rochdale Pioneers' Consumer Cooperative Society, 1844*) in U.K., it was the credit which was more emphasised in the initial stages of the cooperative movement in India. Besides the efforts made by Nicholson in 1897 to popularise the credit cooperatives in India, the Famine Commission in 1901 strongly recommended for the establishment of cooperatives to provide loans to the farmers to improve agriculture.

On the basis of the recommendation of the Famine Commission, the first Cooperative Societies' Act was passed in 1904 thereby formally inaugurating the cooperative movement in India. The 1904 Act provided for the registration of primary credit societies only.

The first step to popularise the non-credit societies was taken in 1912 when the Cooperative Societies' Act was passed to remedy the defects of 1904 Act. The 1912 Act was the first Act passed in India to support the formation of cooperative marketing societies. As per the provisions of the 1912 Act, the first marketing cooperative society i.e. The Kumbakonam Agricultural Society was established in 1913 for supply of seeds, manures and sale of the produce.

Further, Maclagan Committee in 1915 recommended for promoting the non-credit societies. Thus, the period 1904-1915 is regarded as the period of initial efforts and planning for the cooperative movement in India.

The Government of India in 1919 passed an Act to make cooperation a provincial/state subject. The Act gave scope for retaining or modifying the provisions of the 1912 Act. As per the provisions, many provincial acts were passed viz. Bombay cooperative societies Act 1925, Madras and Bihar Cooperative Societies Act, 1932, Orissa Act, 1935, Coorg Act, 1937 and 1941 Act in Bengal. Though the 1919 Act gave stimulus to the cooperative movement in India, it resulted in unplanned expansion of cooperatives. Besides, the great depression of the 1920s gave a tremendous set back to cooperative activity.

In 1927 the Royal Commission on Agriculture (RCA) observed that the main cause of failure of cooperative movement was the lack of requisite education and adequate supervision and guidance. It also observed "if cooperation fails there will be failure of the best hopes of Rural India". Efforts were directed during this time for rehabilitation, reconstruction and reorganisation of the existing societies rather than rapid expansion.

Recognising the neglect of marketing and emphasis only on credit, the Central Banking Inquiry Committee (1929) observed "the emphasis on credit cooperation has still continued and it was not recognised that the question of organised marketing is of greater importance than that of rural credit. To supply the credit requirements of the ryots is important but it is not so much so as it is to cure the disease from which the indebtedness arises".

The first Fruit and Vegetable Growers' Society, The Southern Gujarat Fruit and Vegetable Growers' Association, was established in Pardi in 1935.

The Cooperative Planning Committee was appointed in 1945 and it recommended that the primary credit societies should be converted into multipurpose societies and efforts should be made to bring 30 per cent of the rural population and 50 per cent of the villages within the ambit of reorganised societies within a period of 10 years. It recommended that there should be one cooperative marketing society for every 200 villages and they should handle at least 10 per cent of the marketable surplus within 10 years.

Post Independence Development

During the First Five Year Plan, cooperation was recognised as an instrument of planned economic action in democracy. The plan stressed the importance of progressive realisation of agricultural marketing and processing industries in rural area through the agency of cooperatives. Need for developing cooperative marketing was emphasised so as to ensure better prices for agricultural produce.

The All-India Rural Credit Survey Committee (AIRCSC) was appointed by the Government of India in 1951 under the chairmanship of Mr. Gorwala to look into the causes of failures of cooperatives. The committee submitted the report in 1954. The committee observed that the cooperative marketing system was highly unsatisfactory and only 1 per cent of the marketable surplus was passing through these societies. After analysing the various causes of failures of cooperatives the committee made the famous remark "Cooperation has failed but cooperation must succeed". The committee identified the functional, structural and administrative defects of the credit cooperative structure and recommended for a basic approach i.e. to meet not only credit needs but also the other interrelated economic activities. The committee recommended for an integrated rural credit scheme emphasising the need for state participation at different levels, coordination of credit with other economic activities especially processing and marketing and administration through trained manpower. It also recommended for setting up of a statutory corporation for spearheading the rural economic activities like marketing and processing (NCDC was established on the basis of this recommendation which we will study later).

In 1955, the State Ministers' Conference on Cooperation recommended that cooperative marketing societies should be established at all the assembly markets.

During the Second Five Year Plan large scale development of cooperative marketing, storage and processing was observed as it was recognised that cooperatives have to play a significant role in respect of agricultural production, marketing, cottage and processing industries and international trade.

To strengthen the cooperative marketing network in India, the National Agricultural Cooperative Marketing Federation of India (NAFED) was set up in 1958 as the apex level cooperative.

In 1959, National Development Council (NDC) was constituted and it emphasised the basic objective of cooperative policy as one of rebuilding the rural economy and in particular increasing the agricultural production. It also recommended that marketing should be taken up on a large scale and the existing marketing societies are to be strengthened. The "linking of credit with marketing" was emphasised during this time.

During the Third Five Year Plan, development of cooperative marketing was given a place of special importance in the scheme of integrated rural credit recommended by the Rural Survey Committee. During this period emphasis on formation of marketing cooperatives in general and that of fruit and vegetable in particular was given. Fruit and vegetable marketing cooperatives along with other non-credit societies were established. During the period 1961-66, 22 fruit and vegetable marketing cooperatives came into existence.

The NCDC was established in 1963.

In 1964, an expert committee was appointed under the chairmanship of Prof. Dantwala to review the present pattern of cooperative marketing of agricultural produce. The committee recommended for—

(i) affiliating all agricultural credit societies in the area to a primary marketing society,

(ii) two-tier marketing society i.e. state level apex society and primary society at the mandi level,

(iii) marketing cooperative to undertake both agricultural marketing and distribution and supply function.

By December 1968, there were 35 Fruit and Vegetable processing units. During 1967-68 these units processed 28,298 quintals of fruits and vegetables valued at Rs. 47 lakh. NAFED established a modern multicommodity fruit and vegetable cold storage unit in Delhi.

During the Fourth Five Year Plan various types of cooperative organisations came into existence and the existing ones made good progress.

All India Rural Credit Review Committee (AIRCRC) (*Venkatappaiah Committee*) submitted its report in 1969 and it recommended for the establishment of Small Farmers' Development Authority (SFDA), Agricultural Credit Board, Rural Electrification Corporation, Agricultural Refinance Corporation (ARC) and various institutional agencies for servicing agricultural production programmes including cooperative processing and marketing.

During the Fourth Five Year Plan, for meeting the credit requirements and also to safeguard the farm incomes, emphasis was given to initiate programmes for regulation of markets and strengthening and expansion of cooperative marketing and processing facilities. Further, the effective linking of credit with cooperative marketing was given utmost importance. It was proposed to strengthen the structure of cooperative marketing by establishing 450 new primary marketing cooperatives so as to have a cooperative in every important marketing centre in the country by the end of the Fourth Plan. As part of the programme for the development of cooperative marketing NCDC has initiated various schemes such as establishing technical cells in apex marketing societies, creation of a pool of key marketing personnel, establishment of price fluctuation funds etc. During the plan period it was envisaged to develop fruit and vegetable processing units besides paddy and groundnut.

The main emphasis in the Fifth Plan was on the consolidation and strengthening of the existing primary cooperative marketing societies. Need was felt to organise separate fruit and vegetable marketing societies during this plan. These societies and other specialised commodity cooperatives were envisaged to be organised in area where the concentration of such commodities is more. It was envisaged that by the end of the plan cooperatives would handle agricultural produce worth Rs. 900 crore. In the field of exports cooperatives were expected to step up their operations to Rs. 15 crores by 1978-79. It was proposed to set up 60 cold storages in the cooperative sector for storing fruits and vegetables. Special attention was given during the plan to reduce the inter-state disparities to accelerate the pace of cooperative marketing activities in the weaker areas. A new central sector scheme for the development of the programmes of cooperative marketing, processing, supplies and storage in the weaker areas was proposed during the plan.

In 1977, the Conference of the State Ministers of Cooperation has adopted a policy resolution which set directions for the cooperative movement. The highlights of the resolution are:

(i) Cooperatives to be built up as one of the major instruments of decentralised, labour intensive and rural oriented economic development,

(ii) Cooperatives at all levels shall be closely associated with the process planning for economic development and social change,

(iii) Cooperative movement shall be developed as a "shield for the weak",

(iv) Cooperative movement shall be promoted on a national basis and regional imbalances in cooperative development shall be progressively removed,

(v) Cooperative movement shall be built up as an autonomous, self reliant movement,

(vi) A network of cooperative agro-processing and industrial units shall be built up to provide gainful and economic links between growers and consumers.

The cooperative marketing infrastructure as on 30th June 1976 comprised 3174 primary agricultural marketing cooperatives covering nearly all the important secondary markets in the country, 206 sugarcane supply societies, 172 district level marketing societies, 27 state level marketing federations and NAFED besides 7 commodity cooperative marketing federations. Cooperatives also entered and made significant progress in exports. The value of agricultural produce exported by cooperatives increased from Rs. 26 crore in 1973-74 (last year of the Fourth Plan) to Rs. 148 crore by 1975-76. Marketing societies at the primary level continued to be devilled by a number of problems. The cooperative cold storage capacity increased from 32 lakh tonnes in 1973-74 to 43 lakh tonnes in 1977-78.

During the Fifth Plan the manufacturing and processing in the cooperative sector moved ahead in sugar, oilseeds, fruits and vegetables, cold storage, rice, *dal*, lac, sericulture, cotton, jute, fertiliser, spinning, dairying, poultry, fisheries etc. The scope for establishing marketing and processing cooperative societies for horticultural crops in the North-Eastern region was ascertained during the Sixth Plan.

During the Sixth Plan cooperative movement made some progress in terms of reorganisation of primary agricultural credit societies into strong and viable multipurpose units so that farmers, artisans and self employed get all the services at one point; professionalisation of cooperative management; simplification of procedures for advancing loans. In order to provide a comprehensive service to the rural areas a dynamic and integrated programme of agro-based marketing and processing would be taken up during the plan period. Considerable improvement in management techniques and professionalisation of the same was proposed during this time.

By the end of the Seventh Plan i.e. 1991-92 the cooperatives could handle agricultural produce worth Rs. 7130 crores and 239 cold storages in the cooperative sector were functioning which are proposed to be increased to 309 by the end of the Eighth Plan.

4

PRESENT STATUS OF COOPERATIVE MARKETING OF FRUITS AND VEGETABLES

The present status is analysed in terms of: (i) progress made by the fruit and vegetable cooperative marketing societies, (ii) infrastructure available for cooperative marketing of fruits and vegetables, and (iii) export of fruits and vegetables through cooperatives.

Progress of Fruit and Vegetable Marketing Cooperatives

Trend in Number of Fruit and Vegetable Marketing Cooperatives

There were about 455 Fruits and Vegetables cooperative marketing societies in the beginning of the 1980s (1980-81) which formed about 12 per cent of the total number (3789) of cooperatives engaged in marketing of the agricultural produce. Though there has been an increase in the number of fruit and vegetable marketing cooperatives during the last decade, as in 1990-91 there were about 1190 primary cooperatives besides one National Federation (NAFED) for marketing of fruits and vegetables (*Rajagopal, 1996*), it was very low compared to the total agricultural cooperatives and now the number of fruits and vegetables cooperatives forms about 14 per cent of the total number of societies (8263) for marketing of agricultural produce. Presently, the fruit and vegetable marketing cooperatives are able to handle hardly 1.29 per cent* of the total production of fruits and vegetables. However, since the production of fruits and vegetables is likely to go up due to the importance given in the Eghth Plan, the cooperatives will have to play an active role and more number of marketing cooperatives need to be organised to handle the additional production of fruits and vegetables.

*Due to nonavailability of data on quantity of fruits and vegetables marketed by cooperatives, the value was converted into average quantity at the rate of Rs. 1000 per tonne.

Trend in Value of Fruits and Vegetables Handled by Cooperatives

With the increase in the number of fruit and vegetable marketing cooperatives, the value of fruit and vegetable handled by these cooperatives has increased from Rs. 6900 lakh in 1980-81 to Rs. 43,790 lakhs in 1994-95. However, when compared with the total value of agricultural produce the proportion of F&V has declined from 4 per cent to less than 2 per cent during the last decade. This is perhaps due to the phenomenal increase (almost two fold) in the number and a three-fold increase in the value of agricultural produce marketed by them (Table 4.1). In 1991-92, however, the value of fruits and vegetables handled by cooperatives increased substantially and the proportion of fruits and vegetables in total agricultural produce rose to 4.25 per cent. The main items of agricultural produce have been foodgrains, sugarcane, cotton and oilseeds which formed about 87 per cent of the total produce handled by them..

Table 4.1 : Value of Fruits and Vegetables Marketed by Cooperative Societies

(Rs. in lakh)

Year	*Value of Fruits and Vegetables*	*Value of Agri. produce*	*Fruits & Vegetables as % of Agri. produce*
1980-81	6900	195055	3.50
1981-82	7200	210000	3.43
1982-83	8296	231788	3.56
1983-84	10193	256642	3.97
1984-85	10486	303225	3.46
1985-86	10786	419355	2.58
1986-87	6948	401382	1.73
1987-88	7133	390175	1.83
1988-89	8349	541565	1.54
1989-90	6950	627424	1.11
1990-91	9839	597445	1.65
1991-92	27663	650220	4.25
1992-93	—	683526	—
1993-94	36211	—	—
1994-95	43790	—	—

Source : NCDC Annual Reports.

Performance in Different States

The analysis of the share of different state fruits and vegetables marketing cooperatives and the national agency has revealed that NAFED accounted for over 50 per cent of the total value of fruits and vegetables handled by cooperatives (Table 4.2). The share of NAFED increased from 51 per cent in 1985-86 to around 61 per cent in 1994-95 though there is an absolute increase in the value of fruits and vegetables marketed by cooperatives in all the major fruits and vegetables producing states. The performance of the cooperatives, however, differed from state to state. In Maharashtra, cooperatives like the Ratnagiri District Purchase and Sale Union, Banana Sale Societies in Jalgaon, Grape Sale Societies in Sangli, Vegetable Sale Societies in Ahmednagar and Mango Sale societies in Ratnagiri district performed well. In Gujarat the State Cooperative Fruit and Vegetable Marketing Federation and the Godat Multipurpose Cooperative Farmers' Society did well. In Karnataka the

Table 4.2 : Statewise Value of Fruits and Vegetables Marketed by Cooperatives

(Rs. lakh)

States	*Value of Fruits and Vegetables*			
	1985-86		*1994-95*	
A.P.	46.64	(0.43)	122.2	
Assam	0.49		0.07	
Bihar	54.46	(0.5)	15	
Gujarat	1705	(15.81)	5242	(11.97)
Haryana	2.6		14	
H.P.	283.59	(2.26)	466	(1.06)
J & K			8	
Karnataka	273	(2.53)	370	(8.46)
Kerala	220.85	(2.05)	680	(1.55)
M.P.	18		44.56	
Maharashtra	1026.5	(9.52)	5000	(11.42)
Manipur	1.32		7.2	
Meghalaya	0.29		5.78	
Punjab	56.51	(0.52)	14	
Rajasthan	38.56	(0.36)	36.12	
Sikkim	11.39		15.55	
Tamil Nadu	1438.44	(13.34)	1588.51	(3.63)
Tripura	1.50		2.50	
U.P.	30.75		0.17	
W. Bengal	26.12		65.95	
NAFED, Delhi	5589.95	(51.83)	26687.24	(60.94)
Total	10785.94	(100)	43790.43	(100)

Note : Figures in parentheses are percentages to total.

Source : 1. NCDC Annual Reports.
2. NABARD, Statistical Statements relating to cooperative movement in India (Non-credit Societies).

efforts of HOPCOMS was singly responsible for the commendable performance of the state. In Tamil Nadu, though the Nilgiris' Cooperative Marketing Society (potato), Kodaikanal Cooperative Marketing Society (Hill Banana, Orange, Grapes and Tomato) and the Nilgiris Vegetable Growers' Cooperative Marketing Society, Udhakamandalam have been performing well, the other district societies did not show much progress and hence the value of fruits and vegetables handled by cooperatives in the state increased only marginally from Rs. 1438 lakh to Rs. 1589 lakh between 1985-86 and 1994-95. Cooperatives could not make much progress in other states due to lack of infrastructural facilities.

Infrastructure Facilities for Marketing of Fruits and Vegetables through Cooperatives

Cooperative Cold Storages

The cold storage is an essential part of an efficient marketing system for perishables like fruits and vegetables. However, at present over 90 per cent of the cold storage capacity is in the private sector and such a facility is not available to the primary producers and the available capacity is utilised for storing potatoes only. Mishra (1972) observed that in 1966 about 85 per cent of the total cold storage capacity was utilised for storing potatoes alone. The four major potato growing states of U.P., West Bengal, Bihar and Punjab accounted for 75 per cent of the installed capacity. He also observed that all the cooperative cold storages except the one with NAFED were designed to store potatoes only. Bhatnagar (1985) observed that only 5.20 per cent of the total cold storage capacity was utilised for storing fruits and vegetables. About 92 per cent of the capacity was used for storage of potatoes and the major potato growing states like Punjab, Bihar, and Haryana used about 87 per cent of the installed capacity (Bhatnagar, 1985). Further, it may be noted from Table 4.3 that in 1991, potatoes accounted for about 90 per cent of the capacity followed by multipurpose (8.6%), fish and meat (0.8%), milk and milk products (0.7%).

Fruits and vegetables accounted for less than 1 per cent of the total storage capacity. It was also observed that the private cold storage were mostly used by the private traders and primary producers were able to get this facility only from the cooperative cold stores (*Mohandoss et al.*, 1979). Realising the felt need of the primary producers for storing fruits and vegetables, the National Commission on Agriculture (*NCA, 1976*) recommended for the establishment of cold storages in the co-operative sector and accordingly the National Cooperative Development Corporation (NCDC) started organising cooperative cold stores.

Table 4.3 : Commoditywise Distribution of Cold Storages in India

(As on 31-3-1991)

Commodity	*No.*	*Percent*	*Capacity (cubic meters)*	*Percent*
Potatoes	1820	63.33	20986816	89.60
Multipurpose	416	14.47	1999981	8.54
Fruits and Vegetables	38	1.32	78232	0.33
Meat	17	—	13817	—
Fish	248	8.60	126326	0.54
Meat and Fish	31	—	42531	0.18
Milk, Products	204	7.10	158313	0.67
Others	99	3.44	17880	0.10
Total	2873	100.00	23423896 (7683039t)	100.00

Source : *Agricultural Marketing*, Oct-Dec. 1992.

As seen in Table 4.4, there has been a considerable increase both in number and capacity of the cooperative cold storages. Between 1981-82 and 1994-95, the number of cooperative cold storages increased from 136 to 232 and the capacity of the cold storages in the cooperative sector increased from 2.81 lakh tonnes to 6.60 lakh tonnes. The cooperative sector which accounted for about 5.74 per cent of the total number and 6.42 per cent of the total capacity in 1981-82 has increased its share to 7.78 per cent and 8.41 per cent respectively in 1992-93, the absolute increase being two fold in both number and capacity of the cooperative storages.

However, despite this increase in both number and capacity, the cooperative sector has not been able to help the F&V producers to the extent desired as it accounts for hardly 8 per cent of the total capacity and about 90 per cent of the total available capacity is still utilised for storing potatoes alone with less than 1 per cent capacity being utilised for storing fruits and vegetables. It may be noted from Table 4.5 that the major potato growing states like Uttar Pradesh, West Bengal, Punjab and Bihar have more number of cold stores with higher capacities. However, the major fruits and vegetables growing states like Maharashtra, Gujarat, Tamil Nadu, Karnataka, Himachal Pradesh, Jammu and Kashmir are not equipped with this facility to the desired extent.

Table 4.4 : Growth of Cooperative Cold Storages in India

Year	*Number*	*Capacity (in lakh tonnes)*
1981-82	136 (5.74)	2.81 (6.42)
1982-83	136 (5.54)	2.81 (6.07)
1983-84	144 (5.77)	3.11 (6.45)
1984-85	161 (6.47)	3.71 (7.37)
1985-86	185 (7.34)	4.62 (9.06)
1986-87	204 (7.83)	5.38 (9.96)
1987-88	216 (8.12)	5.85 (10.49)
1988-89	222 (8.08)	6.09 (9.92)
1989-90	229 (8.19)	6.35 (9.35)
1990-91	230 (7.82)	6.43 (8.32)
1991-92	231 (—)	6.51 (—)
1992-93	231 (7.78)	6.55 (8.41)
1993-94	232 (—)	6.60 (—)
1994-95	232 (—)	6.60 (—)

Note : Figures in parentheses are percentages to total.
Source : 1. *Agricultural Marketing*, Oct-Dec., 1992, p. 73.
2. NCDC Annual Reports.

Transport Arrangements

A cost effective transport is an important facility required in proper marketing of fruits and vegetables. The studies on marketing of fruits and vegetables through cooperatives have revealed that the members of the cooperatives could get their fruits transported to the market at a lower cost than the non-members (*Nawadkar et al., 1990*) and in marketing of mangoes, efficient transport in fruit and vegetable cooperatives resulted in the lowest marketing cost of Rs. 53 per crate compared to Rs. 76 per crate in case of other agencies (*Wadkar et al., 1994*). It was reported (*Tilekar et al., 1993b*) that marketing through cooperatives ensured transport right from farm till market, market destination while in case of sale through private traders/commission agents additional responsibility and hence the cost of taking the produce from farm to transport point rested with the producer. However, despite these advantages of transport through cooperatives, refrigerated transport is not available with many of the societies. This facility is essential since ordinary transport results in transit losses and hence increase in marketing cost (*Subrahmanyam, 1986a*).

Table 4.5 : Statewise Position of Cooperative Cold Storages

(As on 31-3-1995)

States	*No*	*Capacity (tonnes)*
A.P.	1	1000
Assam	1	1000
Bihar	20	59850 (4000)
Gujarat	3	3200 (1800)
Haryana	4	12000
H.P.	1	1000
J & K	3	3400
Karnataka	3	2800
M.P.	14	33000 (18250)
Maharashtra	2	1000
Nagaland	1	1000
Orissa	18	23670 (4000)
Punjab	16	22300 (2000)
Rajasthan	3	6000
Tamil Nadu	1	1250
Tripura	1	2000
U.P.	95	282600 (2000)
West Bengal	43	161200 (5800)
Chandigarh	1	1000
NAFED, Delhi	1	2500
Total	232	621770 (37850) 659620

Note : Figures in parentheses show Capacity expanded.

Source : 1. NCDC Annual Reports.

2. NABARD, Statistical Statements relating to Cooperative Movements (Non-Credit Societies).

Cooperative Processing Units

Under the project 'Post Harvest Management of Horticultural Crops' of the National Horticulture Board (NHB), the NCDC provides financial assistance for setting up of the processing units. Under the centrally and corporation sponsored schemes, in 1992-93 it has sanctioned Rs. 190.98 lakh for cooperative fruit and vegetable processing units and it has so far assisted the cooperatives to organise 33 processing units of which 30 are already installed. During 1991-92, these processing units have produced 2700 tonnes of processed products and marketed products worth Rs. 3.02 crore.

Packaging of Fruits and Vegetables

At present fruits and vegetables are packed in bamboo boxes or gunny bags which result in huge losses. With improvements in packaging, plastic trays/crates, corrugated fibre board boxes/trays and plastic film bases are found useful in maintaining quality and reducing the transit losses. The Himachal Pradesh Horticultural Produce Marketing Corporation (HPMC) has recently come out with plastic crates for transporting apples and oranges (*Roy and Pal, 1993; Mehta, 1989; Ranganna et al., 1997*). Such packaging materials, if made available to the Fruit and Vegetable Marketing Cooperatives, can save costs of procurement and distribution of fruits and vegetables through reduction in losses.

Export of Fruits and Vegetables through Cooperatives

The NAFED has been the canalising agent for export of onions and import of fresh fruits. In 1994-95 it exported 4.9 lakh tonnes of onions worth Rs. 256.75 crore and 4631 tonnes of fresh fruits and vegetables valued at Rs. 537 crore. The share of horticulture produce in the foreign trade of NAFED has been around 87 per cent in 1994-95 (Table 4.6) and accounted for about 34 per cent of the total export of fruits and vegetables from India.

Table 4.6 : Export of Fruits and Vegetables by NAFED

(Rs. in lakh

Year	*Fruit & Vegetable export by NAFED*	*Total foreign trade of NAFED*	*Fruit & Vegetable as % of total*
1984-85	6124.79	9150.20	66.94
1985-86	5350.49	7941.71	66.37
1986-87	5708.37	7356.84	77.59
1987-88	4468.62	6627.09	67.43
1988-89	6551.44	7570.39	86.54
1989-90	8464.80	9653.51	87.26
1990-91	11629.04	14273.25	81.47
1991-92	16424.69	22418.81	73.26
1992-93	16884.35	20182.34	83.66
1993-94	25300.30	32798.02	77.14
1994-95	26200.49	29892.00	87.65

Source : Annual reports of NAFED.

During 1994-95, HOPCOMS, Bangalore, MAHAGRAPES, Pune and 4 other cooperatives exported fresh fruits like grapes, pomegranates, tomatoes and strawberry worth Rs. 2.50 crore. With the financial and technical assistance of NCDC, two cooperatives i.e. Markfed Canneries, Jalandhar and Prakash Cooperative Agro-Industrial Society, Sangli exported processed products like mango pulp, banana pulp and canned curried vegetables worth Rs. 1.98 crore to different countries (NCDC, 1995).

Market Intervention and Price Support for Fruits and Vegetables by Cooperatives

Apart from being the canalising agent for export of onions and import of fresh fruits, NAFED has been entrusted with the responsibility of undertaking the Market Intervention Scheme (MIS) and Price Support Scheme (PSS). Under these schemes NAFED has been procuring potato, onion, grapes and kinnow from the major growing states of UP, West Bengal, HP, Punjab, Gujarat, Maharashtra, Haryana and Rajasthan. The details of the price offered, quantity procured from different states are presented in Table 4.7. In 1991-92, NAFED offered a support price of

Table 4.7 : Procurement of Fruits and Vegetables by NAFED under MIS and PSS

Fruits & Vegetables	*Year*	*Support price (Rs/q)*	*Quantity procured*	*Value (Rs. lakh)*	*Major states*
Potato	1984-85	50	17346	98.57	UP, W. Bengal, HP, Punjab,
	1987-88	60-70	3334	23.00	UP
	1988-89 (PSS-89)	65-75	6603	51.34	UP
	1989-90	70-80			
Onion	1988-89 (MIS-89)	60	32460	186.72	Maharashtra
	1989-90 (MIS-90)	60-70	61948	433.94	Gujarat, Maharashtra
	1991-91	70			
	1991-92 (MIS-91)	70-75	4500	33.31	Gujarat
	1992-93	75			
Grapes	1990-91	250B	1256	138.14	Gujarat
		300A	416	10.88	Haryana, Punjab
	1991-92	300B			
		300A	102	3.22	Haryana
Kinnow	1990-91	300A 250B	6585	181.71	Rajasthan, Punjab, HP
	1991-92	300A 250B	2989	86.81	Punjab, Haryana
	1992-93	325A	1703	46.88	Punjab, Haryana
	1993-94	350A 300B 250C	3133	49.49	HP, Haryana, UP

Source : Annual Reports, NAFED.

Rs. 70-75 per quintal for onion procured from Gujarat under MIS. It offered a support price of Rs. 75 per quintal for onion in 1992-93. During 1988-89 NAFED procured potato from U.P. at the rate of Rs. 65-75 per quintal under PSS. A price support of Rs. 300-350 per quintal was offered for grapes procured from Haryana in 1991-92. Similarly, in 1993-94, a price of Rs. 250-350 per quintal (depending upon the grades-A, B, C) was offered for kinnow procured from Punjab, Haryana, H.P. and U.P.

PART III

INSTITUTIONAL SUPPORT

5

THE NATIONAL COOPERATIVE DEVELOPMENT CORPORATION

The National Cooperative Development Corporation came into existence as a result of the recommendation of the Report of All India Rural Credit Survey Committee (1954). The committee recommended that while the RBI wing, now NABARD, shall take upon itself the task of promotion, financing of rural credit cooperatives, the Government of India should set up a statutory corporation for spearheading the development of other rural economic activities like processing and marketing in the cooperative sector. It was with this background that Cooperative Development and Warehousing Board was set up in 1956. In 1963 the warehousing activity was separated and NCDC was created by an Act of Parliament for planning, promoting and financing certain specified rural economic activities on cooperative principles. The NCDC Act was further amended in 1974 to pave way for broadbasing its constitution, resources and for diversification in its objectives and activities on the basis of the recommendation of an expert committee of Government of India. NCDC's main function are to plan, promote and finance programmes of agricultural inputs, processing, storage and marketing of agricultural produce as also supply of consumer goods in rural areas. Over the years, NCDC has emerged as the developmental financing institution for cooperatives.

The NCDC has been the main agency for financing the Fruit and Vegetable Marketing Cooperatives under different schemes.

Strengthening the Share Capital Base and Expansion of Fruit and Vegetable Marketing Activities

The NCDC provides financial assistance upto Rs. 5 lakh for strengthening the share capital base of cooperatives and margin money is provided for improving the working capital base of the societies. Under the corporation sponsored scheme, the NCDC provides financial assistance for purchase of transport vehicles and construction of godowns. Under NCDC III, loan is given for expansion of fruit and vegetable marketing activities.

It may be seen from Table 5.1 that during 1994-95 NCDC sanctioned Rs. 6.48 crore and released Rs. 6.31 crore for marketing of fruits and vegetables. During 1995-96, NCDC assisted 20 new cooperative societies/projects involving assistance of Rs. 94.60 lakh (*Singh*, 1996).

Table 5.1 : Financial Assistance by NCDC for Marketing of Fruits and Vegetables during 1994-95

Sl. No.	*Schemes*	*Amount (Rs. lakh)*	
		Sanctioned	*Released*
A.	Centrally sponsored	26.06	80.53
B.	Corporation sponsored	622.05	530.36
	Total	648.11	630.89

Source : NCDC Annual Report, 1994-95.

During 1992-93 NCDC approved setting up of 7 fruit and vegetable marketing projects in the states of Maharashtra (4), Bihar (1), Madhya Pradesh (1) and West Bengal (1).

Cooperative Cold Storages

As per the recommendations of NCA (1976) NCDC has been giving financial assistance for construction of cold storages in cooperative sector. Almost the entire cold storage capacity in the cooperative sector has been created with the financial and technical assistance of NCDC and the capacity is being utilised by the small and marginal farmers. NCDC has so far released Rs. 71.83 crore for the purpose. As may be seen from Table 5.2, during 1994-95 a total assistance of Rs. 686.23 lakh (Rs. 522.13 lakh as loan and a subsidy of Rs. 164.10 lakh was sanctioned, of which Rs. 309.92 lakh (Rs. 231.92 lakh as loan and Rs. 78 lakh as subsidy) was under centrally sponsored scheme. This was sanctioned for the establishment of 2 multicommodity cold storages of 5000 tonnes each in Bhubaneshwar and Rayagada (Koraput) and for conversion of diffuser cooling system to bunker cooling system of existing 4000 tonnes cold storage in West Bengal. The remaining financial assistance of Rs. 376.31 lakh was sanctioned under the corporation sponsored scheme for: (1) establishment of a 5000 tonnes multicommodity cold storage and 15 TPD Ice Plant in Indore district, (2) meeting the cost escalation of cold storages of 2500 tonnes capacity each in Koyambedu, Tamil Nadu, and in Kolar, Karnataka, and (3) change in cooling system of existing 3400 tonnes capacity cold storage in Indore district of M.P. A total assistance of Rs. 223.83 lakh (Rs. 204.03 as loan and Rs. 19.80 lakh as subsidy) was released during 1994-95.

Table 5.2 : Financial Assistance by NCDC for Construction of Cooperative Cold Storages

Amount (Rs. lakh)

Sl. No.	*Schemes*	*Sanctioned*		*Released*	
		Loan	*Subsidy*	*Loan*	*Subsidy*
1.	Centrally sponsored	231.92	78.00	78.24	19.80
2.	Corporation sponsored	290.21	86.10	125.80	—
	Total	522.13	164.10	204.84	19.80

Source : NCDC Annual Report, 1994-95.

The corporation has availed of the World Bank assistance under NCDC-II IDA Project for expanding the network of cooperative cold storages and marketing of potatoes. In order to make the cooperative units viable and help the cooperatives to create additional cold storage capacity, the NCDC has liberalised the pattern of assistance. Under this pattern, 90 per cent of the project cost is financed. This consists of 40 per cent term loan to cooperatives and 50 per cent investment loan to be provided by the state government as share capital and the remaining 10 per cent to be borne by the beneficiary society. With a view to accelerate the pace of development relating to the establishment of cold storage and ice plants and their rehabilitation, modernisation and modification by the weaker sections in the cooperatively underdeveloped/least developed states/UTs the pattern of assistance has further been liberalised during 1994-95 Under this pattern, cooperatives in the underdeveloped states/UTs are provided 40 per cent loan, 35 per cent share capital and 17.5 per cent to 20 per cent subsidy of the block cost of a cold storage/ice plant.

Cooperative Processing Units

Under the project 'Post Harvest Management of Horticultural Crops' NCDC provides financial assistance for setting up of the processing units. Under the centrally and corporation sponsored schemes, in 1992-93 it has sanctioned Rs. 190.98 lakh for cooperative Fruits and Vegetables processing units and it has so far assisted the cooperatives to organise 33 processing units of which 30 are already installed. NCDC has sanctioned assistance to cooperatives mostly in the North Eastern region for marketing and processing activity.

Assistance under Technical and Promotional Cell Scheme

Under this scheme, NCDC has sanctioned two posts i.e., General Manager (Finance & Accounts) and General Manager (Quality Control) in MAHAGRAPES which is a partnership cooperative of 29 grape growers' societies.

Financial Assistance to NAFED for Price Support Scheme (PSS) and Market Intervention Scheme (MIS) Operations

During 1994-95 NCDC provided an assistance of Rs. 10.39 crore as margin money to NAFED and by March 1995 NCDC has released Rs. 37.20 crore for undertaking PSS and MIS operations.

6

THE NATIONAL HORTICULTURE BOARD

The National Horticulture Board (NHB) was established in 1984 for promotion and development of horticulture sector to strengthen the market information system and also to provide financial assistance for undertaking marketing of horticultural crops. During the Eghth Plan an allocation of Rs. 2000 million is made for NHB.

The major objectives of the Board are:

1. to encourage and promote development of horticulture industry,
2. to provide technological, financial and other assistance for development of horticulture,
3. to assist and develop infrastructure for post-harvest management,
4. to impart training to farmers and in-service officials,
5. to promote consumption of fruits and vegetables in fresh and processed form,
6. to assist the establishment of farmers' cooperative societies to improve the economic and social status of farmers,
7. to assist and organise Udyan Pandit competition, fruit/vegetable/ flower shows.

However, the major thrust of the Board is to create infrastructure for post-harvest management and marketing.

To achieve the above objectives, various schemes have been implemented by the Board and the programmes are proposed to be continued through the Eighth Five Year Plan. The National Horticulture Board is also financing the Fruit and Vegetable cooperatives in marketing and in reducing the post-harvest losses under the following schemes:

Integrated Project on Management of Post Harvest Infrastructure of Horticultural Crops

Under the scheme assistance is provided to cooperatives, registered farmers' organisations engaged in marketing of horticultural produce. Initially, assistance was being provided in the form of subsidy limited to Rs. 15 lakh/beneficiary. However, due to practical difficulties, the pattern of assistance has been changed from subsidy to soft loan during the Eighth Five Year Plan.

Soft loan upto Rs. one crore per beneficiary at a service charge of 4 per cent is given by the Board. The loan is given with a moratorium of upto 3 years and repayment in another 5 years. The main components of the scheme are:

(i) grading and packing centre,
(ii) retail outlets,
(iii) specialised transport vehicles,
(iv) pre-cooling units and cold storage,
(v) auction platform,
(vi) ripening/curing chamber,
(vii) maturity kits,
(viii) rigid plastic container (subsidy at the rate of Rs. 70 per crate or 50 per cent of the actual cost whichever is less).

The Eighth Plan outlay is Rs. 47 crore.

Though the promoters' contribution should not be less than 20 per cent, it is relaxed in case of cooperatives upto 10 per cent.

The pattern of assistance and the components under this scheme are as below:

Description/ Component	*Existing pattern of assistance*	*Approved pattern of assistance during Eighth Plan*
Eligible organisations	Cooperatives, Regd. Farmers/Voluntary organisations/Public sector organisations	No change
Pattern of assistance	(a) 50 per cent subsidy subject to a limit of Rs. 15 lakh per beneficiary organisation	Interest free loan to the extent of Rs. 50 per cent of the project cost subject to a ceiling of Rs. 1 crore per

(Contd.)

Description/ Component	*Existing pattern of assistance*	*Approved pattern of assistance during Eighth Plan*
	torium upto 5 years.	beneficiary with a moratorium upto 5 years.
	(b) For plastic crates 50 per cent subsidy on actual cost or Rs. 50 per crate whichever is less	50 per cent subsidy of actual cost or Rs. 70 per crate whichever is less

Components	*Assistance (Rs. lakh)* Maximum	*NHB Assistance*
Grading/packing centre	0.50	1.10
Retail outlet (Ordinary)	0.10	0.18
Transport vehicle	1.00	1.70
Short duration cool store	1.00	5.00
Plastic crates (per crate)	Rs. 50	Rs. 70
Retail outlet (Air-conditioned)	—	0.75
Refrigerated van	—	5.00
Cold storage (2500 tonnes capacity)	—	35.00
Mechanical grading/packing/waxing line	—	5.00
Auction platform	—	0.50
Ripening/curing chamber	—	5.00
Maturity kits	—	to be decided
Quality testing equipment	—	on case to case
Improved packaging	—	basis

Development of Marketing of Horticultural Produce through Participation in Soft Loan

The scheme was introduced in 1993-94. A financial assistance of upto 40 per cent of the project cost subject to a maximum of Rs. 1.00 crore at a service charge of 4 per cent per annum is given for integrated projects to cooperatives, public sector organisations, NGOs and corporate sector. The Eighth Plan outlay is Rs. 50 crore.

The components of the scheme are:

(i) building including land on which infrastructure is to be created,

(ii) equipment and facilities,

(iii) payment of technical knowhow,
(iv) working capital (margin), promotional expenditure for commercial launching of the product.

Installation of Juice Vending Machines

With a view to popularising the use of processed products as also to ensure the proper utilisation of B-grade fruits which virtually brings much less returns to the producers the Board has initiated a scheme "Alternative Structure for Marketing of Fruit Juices and Fruit based Beverages" during 1987-88 which is to continue through the Eighth Five Year Plan. Under the scheme, assistance is provided to cooperative/agro-horticultural societies for juice vending machines at the rate of Rs. 25000 per machine with 25 per cent subsidy, 75 per cent loan at the rate of 9 per cent per annum.

The Eighth Plan outlay is Rs. 1.35 crore. So far 20 beneficiaries have availed of the assistance and a total of 1023 machines have been installed.

7

THE AGRICULTURAL AND PROCESSED FOOD PRODUCTS EXPORT DEVELOPMENT AUTHORITY (APEDA)

The Agricultural and Processed Food Products Export Development Authority was created by an Act of Parliament in 1986 and is entrusted with the responsibility of export promotion and development of fruits, vegetables, flowers and 12 other scheduled agricultural products. It is managed by a high level body consisting of members of parliament, planning commission, several ministries of central government, trade and industry representatives.

APEDA has been giving financial assistance for the promotion of exports of horticultural products through schematic lending for the development of infrastructure and promotion of transportation by air. The assistance is given directly to the registered exporters under the following schemes:

Scheme for Development of Infrastructure

Under this scheme financial assistance is given to cooperatives, government and private organisations for establishing infrastructural and post harvest facilities. The scheme has the following components:

(a) Financial assistance upto 25 per cent of the capital cost subject to a ceiling of Rs. 1.5 lakh per beneficiary is given for purchase of specialised transport unit for horticulture.

(b) Financial assistance upto 50 per cent of the capital cost subject to a ceiling of Rs. 5 lakh is given to exporters/cooperative organisations and federations for :

(i) establishing pre-cooling facilities,

(ii) setting up mechanised post harvest handling facilities and sheds for grading, sorting, quality control and packaging,

(iii) establishing Vapour Heat Treatment/Fumigation/ Screening machines for exports,

(iv) establishing cold stores at ports/sea ports for export purposes.

The activity components and the pattern of assistance under the scheme are presented below:

Activity component	Pattern of assistance
(a) Purchase of specialised transport units for meat, horticulture and floriculture sector.	25% of the capital cost subject to a ceiling of Rs. 1.5 lakh per unit.
(b) Assistance to exporters/Co-operatives/ federations for:	
(i) Establishing pre-cooling facilities with proper air handling system	50% of the capital cost subject to a ceiling of Rs. 5 lakh.
(ii) Setting up of mechanised post harvest handling facilities and sheds for grading, sorting quality control and packaging	—do—
(iii) Establishing cold stores at air/sea ports for export purposes	100% grant-in-aid to air/sea port authorities.

Scheme for Promoting Transportation of Horticultural Exports by Air

In order to make our horticultural exports competitive in the international markets, APEDA is giving air freight subsidy to the registered exporters including cooperatives to the extent of 25 per cent of the IATA freight rates or in case of fruits and flowers a subsidy of Rs. 10 per kg for exports to Europe and Canada and Rs. 6 per kg for exports to West and South East Asia is provided. The activity component and the pattern of assistance under the scheme is presented below:

Activity component	Pattern of assistance
Air freight subsidy for the following products;	25% of the IATA freight rate for perishables
(a) *Fresh fruits*	OR
Mangoes (Other than Alphonso), Pomegranate, Chickoo, Ber, Custard, Apple, Lychees, Strawberries, Cherry, Bananas, Grapes	(i) Rs. 10 per kg for exports to Europe (ii) Rs. 6 per kg for exports to West Asia, S.E. Asia and CIS
(b) *Fresh vegetables* Asparagus, Broccoli, Capsicum, Mushroom	-do-
(c) *Floriculture* Flowers, Flower plants Cuttings, Bulbs and tubers and Tissue culture	-do-

Scheme for Assistance to Promote Quality and Quality Control

Upto 50 per cent of the cost subject to a ceiling of Rs. 5 lakh per beneficiary is given to exporters producers' associations etc., for setting up/strengthening of quality control activities and laboratories. An assistance upto 50 per cent of the cost subject to a ceiling of Rs. 2 lakh per beneficiary is provided to exporters and producers for specialised consultancy services towards installing ISO 9000 or other recognised international quality control systems and for developing quality control manuals. The actual activity components and the pattern of assistance given under the scheme are presented below:

	Activity component	*Pattern of assistance*
(a)	Assistance to exporters, producers, trade associations, public institutions etc., for setting up/strengthening of quality control activities and laboratories.	50% of the cost subject to a ceiling of Rs. 5 lakh per beneficiary.
(b)	Assistance to exporters and producers for specialised consultancy services towards installing ISO 9000 or other recognised international quality control systems and for developing quality control manuals	50% of the cost subject to a ceiling of Rs. 2 lakh per beneficiary.

Scheme for Packaging Development

Under this scheme financial assistance upto 60 per cent of the cost of packaging development subject to a ceiling of Rs. 1 lakh per beneficiary is given to the exporters for packaging development. APEDA is also giving a subsidy upto 30 per cent subject to a ceiling of Rs. 1 lakh per beneficiary for utilising the packaging developed by APEDA through Indian Institute of Packaging (IIP) or others along with 'Produce of India' label. The assistance at present is available for packaging for mangoes, grapes, kinnow and apples in approved cartons. The activity components and the pattern of assistance under this scheme are presented on next page.

Scheme for Export Promotion and Market Development

In order to encourage planned and organised effort for export promotion and market development and hence enhanced exports of agricultural and

Activity component	Pattern of assistance
(a) Programme for development of packaging standards and designs	APEDA's internal scheme
(b) Assistance to exporters for packaging development	60% of the cost of packaging development subject to a ceiling of Rs. 1 lakh per beneficiary
(c) Assistance to exporters for utilising packaging developed by APEDA through IIP or others along with "Produce of India" label	30% subsidy subject to a ceiling of Rs. 1 lakh per beneficiary

processed products APEDA is giving financial assistance to the exporter, The scope of the activities and funding under the scheme is presented below:

Activity component	Pattern of assistance
(a) Supply of product samples for test marketing, product information and promotion	Cost of samples or freight or both, to be decided on case to case basis subject to ceiling of Rs. 50,000 per beneficiary.
(b) Publicity and promotion through preparation of product literature, publicity material etc.,	40% of the cost subject to a ceiling of Rs. 1 lakh per beneficiary.
(c) Brand publicity through advertisements	40% of the cost subject to a ceiling of Rs. 50,000 per beneficiary.
(d) Product/market promotion by APEDA	APEDA's internal scheme
(e) Participation in international trade fairs abroad in which APEDA is participating directly.	Assistance on space rentals to be decided on fair to fair basis.

Scheme for organisation building and HRD

In order to upgrade the skills and managerial capabilities, and also create awareness about the export production and marketing techniques specialised training programmes can be organised by the exporters/ growers' associations with the help of financial assistance from APEDA. The scope of activities and the funding under this scheme are as follows:

Activity components	Pattern of assistance
(a) Assistance for upgradation of technical skills of supervisory, technical and managerial personnel through training in India	Upto 50% of the cost of the approved training programme.
(b) Assistance to recognised associations of growers/exporters for information dissemination	50% of the cost of publications for the purpose of information dissemination subject. to a ceiling of Rs. 10.000 per association.

PART IV

CASE STUDIES

8

NEED FOR IN-DEPTH STUDIES OF FRUIT AND VEGETABLE COOPERATIVE MARKETING SOCIETIES

The most important peculiarity of the co-operative organisation is their varied performance, though all of them are supposed to have a basic common principle and organisational structure. For example, the cooperatives are supposed to be formed by a group of people who have a common interest to be taken care and to be governed by the elected representatives of the members. But unfortunately most of the co-operative societies in India, neither protect the interest of the members nor formed by coming together by the people on their own based on common felt need. Adding to this the interference of politicians and vested interest of people have reduced these societies to the status of being the means for exploitation for the benefit of a few individuals.

But the recent success of the Milk co-operative societies organised by the National Dairy Development Board (NDDB), Anand (Gujarat), has given hopes, that the co-operative societies can be successfully organised to take care of the interest of the producer members. Hence a number of co-operative societies for marketing of various commodities including fruits and vegetables have been started in various states. But unfortunately most of these societies have failed. The reason for the Milk co-operative societies to be successful, though it is also a perishable commodity like fruits and vegetables and for the failure of the latter can be known if one looks at the differences more closely. In case of Milk, the reasons for their success are:

1. Milk is considered as an essential commodity and is required daily by all sections of the people i.e. poor, middle class and rich people.

2. The processed products of Milk like milk powder, ghee, curd etc. are also considered as essential commodities and are in great demand.
3. The Milk co-operative societies throughout the country are backed up by their own processing facilities for utilisation of the excess milk.
4. Only one commodity i.e. Milk is handled by these societies.
5. Milk is produced throughout the year and marked seasonal fluctuations are minimum.
6. Milk price is also stable throughout the year without much fluctuations.

In contrast to the above if one looks at the Fruit and Vegetable co-operative societies, the problems faced are:

1. The fruits and vegetables are not considered as important essential items and their consumption is restricted based on the seasonal availability, prices etc.
2. The fruits and vegetables are highly perishable and the post-harvest losses are also high.
3. The processed fruit and vegetable products are also not considered as essential items as in case of milk processed products like curd, ghee etc.
4. The processed fruit and vegetable products are costly and beyond the reach of the common man.
5. Most of the processing facility for fruits and vegetables is in the hands of the private sector.
6. Most of the co-operative societies have to handle more than a single fruit or vegetable each needing different types of handling and disposal.
7. The highly seasonal nature of the fruits and vegetables results in gluts and scarcity which would result in wide price fluctuations making their marketing a risky proposition.
8. Fruits and vegetables are bulky in nature resulting in high transport costs and other problems.

In view of the above differences, it is not basically correct to expect the same type of success in fruit and vegetable co-operative societies as in case of Milk co-operative societies.

Inspite of the above drawbacks, we can find a few fruit and vegetable co-operative marketing societies running successfully, whereas most of the others have failed. Hence to know the reasons for their success and also causes for their failures it is essential to take up an in-depth study of these societies.

For making in-depth studies, it may not be possible to take a large number of these societies and hence generally the case study approach is adopted and the same is discussed in the next chapter.

9

DATA AND METHODOLOGY FOR STUDYING FRUIT AND VEGETABLE MARKETING COOPERATIVES

As stated earlier in Chapter 8 there is a need to make in-depth studies of fruit and vegetable marketing cooperatives to understand the reasons for their success and the causes for their failures. For this purpose either survey method or case study method could be adopted.

Benefits of Case Study Methodology

While the survey method is based on collection of data from a large sample of different societies the case study methodology is based on intensive study of comparatively less number of sample, sometimes confined to even a single case. The case study method helps to secure in-depth details about the unit of the study which provide clues and ideas to know as to what exactly is happening and enables to pinpoint the problems if any as against survey method which result in gathering more quantity rather than quality data.

Considering the above advantages of the case study methodology we adopted case study method for understanding the performance of the fruit and vegetable marketing cooperatives.

Selection of Cases

The case studies are so selected as to highlight both successful and not so successful efforts in marketing of fruits and vegetables by cooperatives. Brief description of the selected Fruit and Vegetable marketing cooperatives is given below:

HOPCOMS, Bangalore

The present HOPCOMS was established as 'The Bangalore Grape Growers' Cooperative Marketing and Processing Society Ltd' (BGGCOMS) in 1959 with the main objective of encouraging grape cultivation by providing the necessary inputs, marketing facilities etc. However, since 1965 the society started handling other fruits and vegetables also and in 1983 the name of the society was changed as 'The Bangalore Horticulture Producers' Cooperative Marketing and Processing Society Ltd. (BHOPCOMS) and it became HOPCOMS in 1987. The HOPCOMS has been considered a successful venture in the field of marketing of fruits and vegetables and many states have shown keen interest to start Fruit and Vegetable Cooperatives on the lines of HOPCOMS.

DKHOPGCOMS

The Dakshina Kannada District Horticulture Produce Cooperative Marketing and Processing Society Ltd was started by the Department of Horticulture, Dakshina Kannada District in 1965. The main objective of the society is to arrange for the sale of horticulture produce of the members, providing short term loans and production requisites to the members and also to make available the modern technical know-how.

FRESH, Hyderabad

The Farmers' Rural Extension Service in Horticulture (FRESH) Marketing Cooperative Society Ltd. was started on the lines of HOPCOMS with NCDC assistance in 1987. The society has not been able to make profits ever since its inception and has not been so successful in effective marketing of fruit and vegetables.

Cooperative Fruit (Banana) Sale Societies, Jalgaon

There are a few cooperatives specialised in marketing of a single commodity like milk and sugarcane. These societies were found successful not only in making profits but also in safeguarding the interests of the members. (Ranade *et al.*, 1988; Gupta 1986) The main reason for their success was single commodity orientation. The fruit and vegetables, unlike other perishables like milk, have their own drawbacks as discussed earlier, which make it difficult to have a single fruit or vegetable cooperative. Despite these drawbacks, we do find a few cooperatives dealing in banana which are found to be successful. Hence two cooperative fruit (banana) sale societies, namely The Kissan Cooperative Fruit Sale Society, Nimbhora, and The Jalgaon Fruit and Vegetable sale society, Jalgaon were selected for detailed study.

Methodology Adopted for Case Studies

Compound Growth Rate

Annual percentage growth rate in membership, share capital, turnover etc. was analysed using the exponential form of the production function as below:

$$y = ab^te^u \qquad \ldots (1)$$

where

y = dependent variable
t = time trend
a = intercept (value of y when $t = 0$)
$b = (1 + r)$, r being the growth rate
u = error term

By transforming equation (1) into semi log form,

$$\log y = \log a + t \log b + u \qquad \ldots (2)$$

Log a and log b are obtained by application of least squares procedure to equation (2) and the growth rate (r) is computed as below:

$$r = (\text{Antilog of } \log b - 1) \times 100$$

Whenever the length of the time series was fairly long, as in case of Jalgaon societies, compound growth rate (%) at constant prices was also worked out to account for inflation.

Ratio Analysis

Ratio analysis is the process of determining and interpreting the numerical relationships based on financial statements. A ratio is a statistical yardstick that provides a measure of relationship between variables or figures (Kuchhal, 1988). In order to test the liquidity, solvency and profitability of the business of the cooperatives, the following accountancy ratios were used. The computation of the ratios is as below:

(i) Current Ratio (CR)

The ratio is also called working capital ratio. The ratio indicates the liquidity of the assets and the extent to which the current assets are available to meet the current liabilities as and when they become due. The ratio is calculated as below:

$$CR = \frac{CA}{CL}$$

where

CA = Current assets which includes cash in hand, cash in bank, marketable securities, sundry debtors (accounts receivable), advances to members, pre-paid expenses and inventory

CL = Current liabilities which includes sundry creditors (accounts payable), short term loan, interest payable etc.

The recommended ratio is 2, but the ratio may vary at different phases of the business.

(ii) Quick Ratio (QR)

This is also called as the acid test ratio and is a better test of financial strength than the current ratio as it gives no consideration to slow moving inventory. It is a supplementary measure of liquidity and places more emphasis on immediate conversion of assets into cash.

The ratio is calculated as

$$QR = \frac{QA}{CL}$$

where

QA = quick assets = (Current assets – inventory)

CL = Current liabilities

The recommended ratio is 1 but may vary at different phases of the business as in case of current ratio.

Ratio of Total Assets to Total Liabilities (TATLR) : As the name indicates this ratio is worked out as :

$$TATLR = \frac{TA}{TL}$$

where

TA = Total assets which include current assets, fixed assets and investments.

TL = Total liabilities which include current liabilities, long term loans, share capital, debentures, outside liability etc.

The recommended ratio should be more than one so as to enable the society to expand its business.

Ratio of Total Liabilities to Owned Fund (TLOFR) : The ratio is worked out as:

$$TLOFR = \frac{TL}{OF}$$

where

TL = total liabilities,

OF = Owned Fund which includes share capital, reserves etc.

The ratio indicates the solvency of the society i.e. whether the society has been able to meet its obligations from the owned fund itself, and if so to what extent. The general rule is that the owned fund must be atleast equal to or greater than the outside investments or borrowed funds.

Ratio of Current Liabilities to Owned Fund (CLOFR) : The ratio is calculated to know as to whether the society has been able to meet its current obligations from the owned fund itself. The ratio is computed as:

$$CLOFR = \frac{CL}{OF}$$

where

CL = current liabilities

OF = owned fund

Ratio of Fixed Assets to Owned Fund (FAOFR) : The ratio is calculated to know the extent of investment of owned fund in fixed assets and it is computed as:

$$FAOFR = \frac{FA}{OF}$$

where

FA = fixed assets which include movable properties like furniture, vehicles less depreciation and immovable properties like land and building less depreciation etc.

OF = owned fund

Though there is no specific recommendation, a high ratio is considered dangerous as it reduces the availability of working capital to the society.

Ratio of Fixed Assets to Total Assets (FATAR) : The ratio is calculated as:

$$FATAR = \frac{FA}{TA}$$

where

FA = fixed assets

TA = total assets

There is no specific recommendation but it is advisable that the business maintains a relatively less fixed assets to total assets ratio so as to make available as much assets as possible in liquid form.

Net Profit to Turnover Ratio (NPTOR) : The ratio is calculated as:

$$NPTOR = \frac{NP}{TO}$$

where

NP = net profit

TO = total turnover

A high ratio indicates management efficiency.

10

CASE STUDY 1

THE HORTICULTURAL PRODUCERS COOPERATIVE MARKETING AND PROCESSING SOCIETY LTD. (HOPCOMS), BANGALORE

In Karnataka, the cooperative movement for marketing of fruits and vegetables started in the mid 60s with the establishment of district level societies, though at present majority of them are either not performing well or not functioning at all. For example, out of 54 fruit and vegetable cooperative marketing societies registered in the state only a few district level societies like Belgaum Gardeners' Cooperative Production Supply and Sale Society Ltd. and the Horticultural Produce Cooperative Marketing and Processing Society, Ltd (HOPCOMS), Bangalore have performed well (GOK, 1992). Even in case of HOPCOMS, the picture was not all that rosy. In fact, till late 70s HOPCOMS was a perpetual loser as it followed certain wrong policies of procurement, sale etc., (*Subrahmanyam et al, 1979*). However, in the 80s, HOPCOMS made a dent in the marketing of fruits and vegetables in Karnataka and HOPCOMS, now, is a success story. In view of these it was felt that a detailed study of performance of HOPCOMS would be useful.

DATA AND METHODOLOGY

Relevant data pertaining to the organisational, business and financial aspects of the society for the period from 1987-88 (when the society became HOPCOMS) to 1992-93 were collected from the records of the society at the head office, Bangalore and the HOPCOMS branch at Mangalore with the help of a specially designed questionnaire. Simple tables and graphs were used as tools of analysis besides certain accountancy ratios.

RESULTS AND DISCUSSION

The results of the analysis of data are discussed under the following heads:

I. Organisation of HOPCOMS,
II. Business performance of HOPCOMS,
III. Financial performance of HOPCOMS.

Organisation of HOPCOMS

The present HOPCOMS was established as 'The Bangalore Grape Growers' Cooperative Marketing and Processing Society Ltd' (BGGCOMS) on 10th September 1959 with the main objective of encouraging grape vine cultivation by providing the required inputs, technical know-how, marketing facilities etc. The society started handling fruits and vegetables apart from grapes since 1965. In 1983, the name of the society was changed as 'The Bangalore Horticultural Producers' Cooperative Marketing and Processing Society Ltd. (BHOPCOMS) and subsequently in 1987 it became HOPCOMS.

Membership and Share Capital

The membership of the society consists of four categories viz. 'A' class members, who are the producers of horticultural crops in the area of operation, 'B' class members, who are admitted as associate members and include cooperative institutions, 'C' class earmarked for the Government of Karnataka, and 'D' class members comprise traders and commission agents.

At the time of establishment of the society, there were 507 members with a paid-up capital of Rs. 12,675. As may be seen from Table 10.1, membership as well as the share amount has increased over the years. In 1970, the membership stood at 1245 and the share capital was Rs. 85,983 with a Government contribution of around 35 per cent and at the end of March 1993, there were 6199 members with a paid-up capital of Rs. 186.22 lakh. It may be noted from the table that despite an increase in the membership, which basically came from 'A' class i.e. producers, the share of the Government increased over the years and it was as high as 95 per cent in 1992-93, whereas the share of the producers remained stagnant at around 20 per cent up to 1990-91 and came down to 5 per cent thereafter, which gives an indication that HOPCOMS is rather a Government organisation than a producers' cooperative society.

Table 10.1 : Growth of Membership, Share Capital and Government Contribution

Year	*Membership (No.)*	*Paid-up capital (Rs. lakh)*	*Government contribution (%)*
1970-71	1245	0.86	35.0
1977-78	1304	2.99	77.0
1987-88	5311	23.97	80.0
1988-89	5426	24.22	78.0
1989-90	5652	26.09	81.0
1990-91	5981	26.44	79.0
1991-92	6156	96.92	94.0
1992-93	6199	186.22	94.5
CGR (%)	3.50	51.00	

The society is authorised to raise share capital worth Rs. 10 crore by issuing 4.9 lakh shares to 'A' class, 10,000 shares to 'B' class and 5 lakh shares to the Government. Each share is valued at Rs. 100.

The jurisdiction of the society extends to 8 districts of Karnataka, namely, Bangalore (both rural and urban), Mysore, Dakshina Kannada, Kolar, Mandya, Tumkur and Shimoga. The society has one branch each in 6 districts barring Shimoga and Bangalore (Rural).

Management

HOPCOMS is run under the guidance of the Department of Horticulture and is managed by a Board of Management consisting of 15 members—11 elected from 'A' class and 4 Government nominees. The Director of Horticulture is the Ex-Officio President of the society. However, since 1992-93, the president is elected from among the 'A' class members. A senior class I officer of the Department of Horticulture is the Managing Director of the society. One Joint Registrar of Cooperative Societies and a Special Officer (Grape Development) from the Department of Horticulture are the other two Government nominees.

Business Performance of HOPCOMS

The main business of HOPCOMS is procuring and disposal of fresh fruits and vegetables and the various channels followed for carrying out this activity is presented in Fig. 10.1 and discussed briefly below:

1. Procurement of Fruits and Vegetables

The society procures fruits and vegetables both from cultivators (members as well as non-members) and the open market as discussed below:

(i) Supply from Cultivators

(a) Supply at the Head Office/Branch : Producers at the nearby places bring their produce on their own and supply at the H.O. or at the branches. The cultivator has to take an indent from the society for the supply of fruits and vegetables and normally, produce in excess of the indented quantity will not be accepted. The society bears the unloading charges and it makes payment to the cultivators immediately after procurement upto Rs. 3000 in cash and if it exceeds Rs. 3000, then cheque is issued to them.

(b) Supply at the Procurement Centres : In the mid 80s, HOPCOMS opened procurement centres at Sarjapur, Hoskote and Doddaballapur and of late, in Hassan and Channapatna. The fruits and vegetables growers in the nearby areas supply their produce at these centres. For transporting this to the H.O., the society charges a transport cost of 10 to 20 paise per kg. of fruits and vegetables.

As can be seen from Table 10.2, in the 70s, the society was procuring hardly 35 to 40 per cent of fruits and vegetables from the field. However, in the 80s, there was a change in the policy of HOPCOMS in favour of field procurement and with the help of the procurement centres, at present, the society purchases nearly 85 per cent of fruits and vegetables from the cultivators directly. Almost entire quantity of tomato, cabbage, cauliflower, cucumber, raw banana, pomegranate, papaya and mango is now being procured from the field.

(ii) Procurement from the Market

A part of the produce is also bought from the local markets to meet the requirements of the bulk buyers like Government hospitals, hostels, factories etc. However, though this helps the society meet its commitments, as is evident from Table 10.3 the society pays a higher

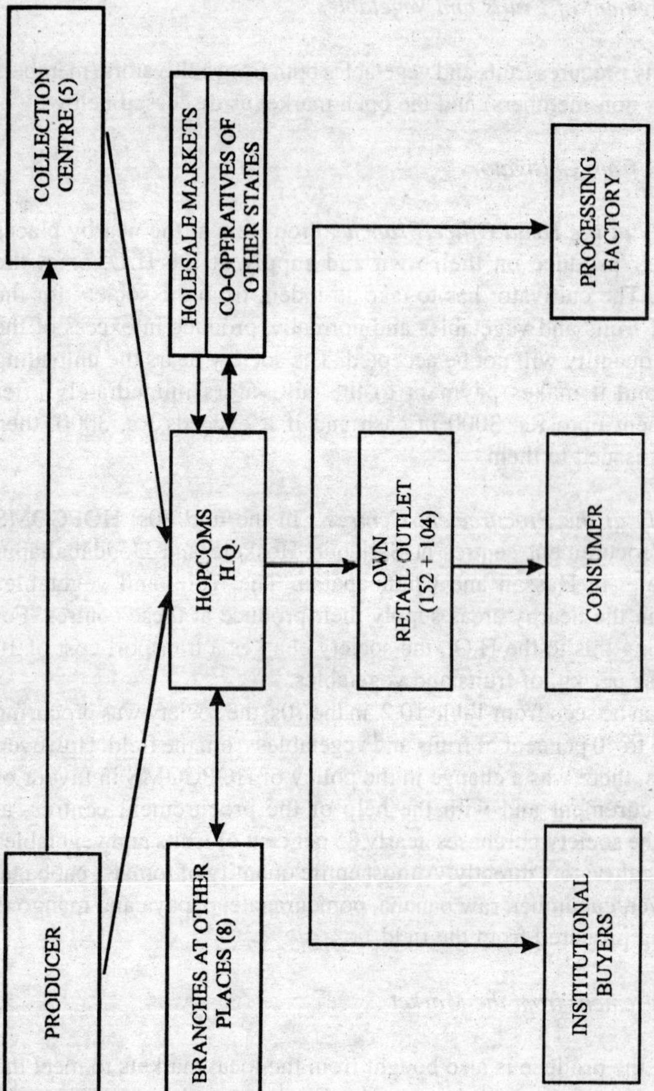

Fig. 10.1 : Marketing Channels of HOPCOMS

Table 10.2 : Share of Field Procurement in Total Procurement of Fruits and Vegetables

(Percentages)

Fruits/Vegetables	*1977-78*	*1992-93*
Vegetables		
Tomato	62.12	100.00
Beans	34.50	98.00
Cabbage	35.00	100.00
Cucumber	15.74	99.00
All vegetables	—	87.00
Fruits		
Banana	75.00	92.00
Mango	16.50	95.00
Papaya	77.50	100.00
All fruits	—	78.00
Total F &V	—	84.00

Source : Subrahmanyam *et al.* (1979) for 1977-78 and HOPCOMS, Bangalore for 1992-93.

price for fruits and vegetables whenever it resorts to market purchases. The price differential is as high as Rs. 4-6 per kg for fruits and around Rs. 1 per kg for vegetables. Thus, on an average, the society loses Rs. 3 per kg of fruits and vegetables by purchasing from the market. It was observed that the policy of buying more from the market followed by the society in the 70s resulted in net losses to the society (*Subrahmanyam et al., 1979*). The society buys about 15-20 per cent of fruits and vegetables from the market. If this is also purchased from the members' field, then the society not only gains Rs. 3 per kg in terms of price but also helps the producers in better marketing of their produce.

Table 10.3 : Price Differential due to Method of Procurement (1992-93)

Fruits/Vegetables	*Price (Rs./kg)*	
	Field	*Market*
Vegetables		
Beans	3.56	4.70
Carrot	3.42	5.18
Cucumber	2.20	2.34
All Vegetables	3.29	3.59
Fruits		
Mango	6.73	9.60
Orange	7.18	10.78
All fruits	5.59	10.71
Total Fruits and Vegetables	3.99	6.80

(iii) Supply from Other States

In addition to procurement from producers and the market, HOPCOMS gets a small quantity of the produce from the other states. It gets apple from NAFED, The Himachal Pradesh Horticultural Produce Marketing and Processing Corporation (HPMC), National Dairy Development Board (NDDB) and GROWREP, Delhi, kinnow orange from GROWREP, orange from NAFED, Nasik and onion from Vegetable and Fruit Cooperative Marketing Society (VEFCO), Nasik. The procurement of fruits and vegetables is made on consignment basis.

(iv) Grading of Fruits and Vegetables

Though HOPCOMS does not classify fruits and vegetables into grades like A, B, C, the society claims that it maintains the quality of fruits and vegetables by accepting only the good quality produce from the growers. It rejects the injured, damaged and diseased ones. Although this helps the society minimise the wastage and hence the loss, yet, from the producers' point of view this is not desirable. In this connection, we feel that HOPCOMS can improve its processing facility for fruits, as has been done in case of 'KANCHAN' (*Gupta and Gupta, 1985*), and buy even the inferior quality fruits from the growers and pay accordingly. In case of vegetables, it can have contractual arrangements with the processors and buy the damaged vegetables also from the cultivators instead of rejecting the entire lot and thereby causing inconvenience to the fruit and vegetable growers, especially those bringing the produce from far off places.

Earlier, the society was deducting 20-30 per cent of the produce of the growers towards driage and wastage while making payments (*Ananth, 1984*). Now it claims that the practice of physical quantity deduction is completely stopped. However, from our discussion with the staff, farmers and from personal observations, we came to know that about 5-6 per cent of cabbage, cauliflower and beans is still deducted towards driage and spoilage. We also observed that in case of cabbage, payment is made only after the entire quantity is sold so as to account for the loss in quantity due to driage.

(v) Indent System

From our earlier discussion, we understand that the society follows the 'indent system', which it used to follow even in the 70s, while procuring fruits and vegetables from the growers. This system, no doubt, helps the society in regulating the supply based on the demand. In fact, we were told that the disposal of the produce is the major problem of the society and hence, in order to avoid losses due to surplus, HOPCOMS resorts to this system. We wonder, however, as to how this system would help the member growers, who are supposed to be the real beneficiaries of the society. This, we feel, is in fact acting as a hindrance to the enhanced participation of the members in the activities of the society. First of all, the membership of 6200 itself is very poor and of these 6200, we learnt that, only 2000-3000 members are regularly supplying fruits and vegetables to the society. The society can dispense with this system by accepting the produce only from the members, which would help regulate the supply. Further, the society can also know the supply from the members by suggesting the vegetable crop plans depending upon its requirement for the bulk buyers.

2. Disposal of the Produce

Sales through Retail Outlets

HOPCOMS has a good network of 256 retail outlets spread over 8 districts. These outlets are run by the salesmen of the society who get a commission of 3.7 per cent from the society. It may be observed from Table 10.4 that the H.O. Bangalore sold about 71 per cent of vegetables and 79 per cent of fruits through these retail outlets. Further, about 80 per cent of vegetables like cowpea, bhendi, knolkhol and tondekai (coccinea) and over 60 per cent of tomato and brinjal were sold through these retail outlets. As regards fruits, around 95 per cent of sapota, papaya, pomegranate, pineapple and banana (*yelakki*) and over 65 per cent of orange, grape and banana reached the consumers through these outlets. It may also be observed that HOPCOMS gets higher price for fruits and slightly less for vegetables when they are sold in these outlets.

Sales to Bulk Buyers

HOPCOMS sells fruits and vegetables on bulk basis to certain 'Institutions' like Government hospitals, hostels, factories and also to processors like KISSAN and Karnataka Agro-Fruits. Normally, HOPCOMS supplies fruits and vegetables on credit basis and from the

Table 10.4 : Share of Different Modes of Sales of Fruits and Vegetables and Price Differential (1992-93)

Fruits & Vegetables	*HOPCOMS Outlets*		*Institutions*	
	Share (%)	*Price (Rs/kg)*	*Share (%)*	*Price (Rs/kg)*
Vegetables				
Tomato	62	1.37	38	1.53
Brinjal	63	4.10	37	3.10
Cabbage	64	3.54	36	3.44
Okra	88	6.01	12	6.53
Cucumber	76	2.96	24	5.52
All vegetables	71	2.67	29	6.43
Fruits				
Apple	56	19.08	44	13.91
Pineapple	98	7.42	2	5.66
Banana (Y)	96	6.26	4	5.06
B.B. Grapes	63	8.00	37	11.00
Pomegranate	94	14.01	6	10.00
All fruits	79	8.67	21	4.93

records we came to know that it charges 40-50 paise per kg of vegetables more than the stall price when vegetables are sold to the factories. In case of processors, transport cost is added to the price of the vegetables. This, perhaps, is the reason for the higher price that HOPCOMS gets for vegetables like tomato, bhendi, cucumber, onion etc. when it sells them to the bulk consumers. This gives us a clue that, as noted earlier, HOPCOMS can think of increasing their sales of vegetables by having contractual arrangements with these bulk consumers.

Sales to HOPCOMS Branches

The society sends fruits and vegetables to its branches to be sold through their retail outlets. The branches, mostly, act as distribution centres of the H.O. As a matter of fact, during our visit to Mangalore branch of HOPCOMS, we observed that nearly 40 per cent of fruits and vegetables procured by the branch comes from the Head Office. The picture may not be different in other branches as well. However, we feel that the branches of HOPCOMS, instead of acting as the mere distribution centres, should function as independent units of fruit and vegetable growers of the respective district. In order to help the fruit and vegetable growers, unlike the Mangalore branch (which has not enrolled any producer as its member even after 5 years of its existence), membership enrolment drive should gain momentum in the branches. They can procure the produce from their members and sell them through retail outlets.

Progress of Sales of Fruits and Vegetables

It may be observed from Table 10.5 that marketing of fruits and vegetables being the major activity of the society, sales of fruits and vegetables account for over 88 per cent of the total sales. It may also be observed that there was more than a two-fold increase in the sales of fruits and vegetables between 1987-88 and 1992-93. This increase was made possible through the increased number of retail outlets over the years. In fact, consequent upon the opening up of the procurement centres and hence the enhanced supply of fruits and vegetables, HOPCOMS opened retail outlets in different districts for their distribution. As a result, the number of these outlets increased from a mere 26 in 1981-82 to 115 in 1987-88 and thereafter to 256 in 1992-93 which helped HOPCOMS increase the sales of fruits and vegetables.

Table 10.5: Progress of Sales of HOPCOMS

(Rs. in lakh)

Year	*Fruits & Vegetables*	*Inputs*	*Juice*	*Total*
1987-88	666.45 (89)	57.56(7)	22.01 (1)	747.14
1988-89	913.25 (89)	77.18 (8)	22.82 (2)	1023.20
1989-90	798.98 (87)	87.12 (10)	13.32 (1)	913.96
1990-91	1202.49 (88)	132.09 (10)	18.03 (1)	1370.67
1991-92	1480.32 (89)	182.94 (11)	20.00 (1)	1704.50
1992-93	1828.39 (90)	152.59 (8)	23.89 (1)	2030.48
CGR (%)	21.81	25.24	0.91	21.94

Note :
1. Figures in parentheses are percentages to total.
2. Total includes sales of ice cream and canteen items also and excludes closing stocks.
3. Figures for 1989-90 are only for 9 months.

Price Policy

HOPCOMS has an approved policy of fixing the procurement price slightly higher than the prevailing wholesale price in the market and the stall (outlet) price at a slightly lower level than the ruling retail price so as to maintain a margin of 25 per cent (This margin was 15 per cent up to 1989-90 and it was increased to 20 per cent in 1990-91 and later on to 25 per cent). An analysis of monthly prices of certain fruits and vegetables for the year 1992-93 revealed that the fruits and vegetables grower gets 75 per cent of the stall price. We also observed that this generally earns more (33 per cent) than the approved margin of 25 per cent.

Benefits to Producers and Consumers

In order to understand as to whether or not the producer/consumer is benefited by the price policy, a comparison is made of the daily prices at the market with that of HOPCOMS for the month of July 1993 in Table 10.6. It may be seen that the price paid to the producer is 6-10 per cent higher than the wholesale market price. This means that the producer gets the full wholesale price and a part of the retail margin. If we account for the commission charges, which is normally 8-10 per cent of the wholesale price itself, then the producer is likely to get 18-20 per cent higher than the price which he gets in the market. As regards the consumer, he is also benefited by the price policy as he gets the same vegetable or fruit at 10-12 per cent less than the retail price in the market.

Table 10.6: Prices of Fruits and Vegetables at HOPCOMS *vis-a-vis* Market (Rs/kg)

	Wholesale Price		*Retail Price*	
	Market	*HOPCOMS*	*Market*	*HOPCOMS*
Vegetables				
Tomato	5.33	5.62	8.00	7.49
Beans	3.13	3.07	4.76	4.09
Cabbage	2.61	2.90	4.47	3.87
Brinjal	3.41	3.72	5.63	4.96
Okra	2.94	3.32	5.48	4.42
Fruits				
Apple	20.89	19.44	28.04	26.11
Pineapple	5.00	5.25	8.10	7.00
Banana (Y)	7.00	7.10	10.95	9.50
Grapes	6.51	6.83	10.32	9.11
Pomegranate	11.22	10.67	17.37	14.22
Mango (Neelam)	9.11	8.57	12.91	11.43

3. Production Related Activities

HOPCOMS supplies production requisites like vegetable seeds, fertilisers, PPC (fungicides and insecticides) and garden implements to the fruits and vegetables growers at reasonable price. It may be observed that inputs account for 8-10 per cent of the total sales of HOPCOMS (Table 10.5). Further, it is also to be noted that there has been a 3 fold increase in the value of inputs supplied to fruits and vegetables growers. This, we believe, is due to the opening up of the fertiliser and PPC godowns at Kolar, Chikkabballapur and Vijayapur.

4. Processing Activity

HOPCOMS takes up preparation of juice from grapes, mango, crange, apple etc., in Bangalore, Mysore and Mangalore branches and sells it in bottles of 200 ml in their retail outlets. Although, with the opening up of the procurement centres, there was an increase in the supply of fruits, as may be seen from Table 10.5, a corresponding increase is not observed in their processing and juice sales have remained at around Rs. 20 lakh (though there was an improvement in 1992-93) accounting for hardly 1 per cent of the total sales of HOPCOMS as specific efforts were not made either in the juice preparation or its sales. Juice has been fetching a higher margin than any other activities of the society. The economics of juice preparation presented in Table 10.7 clearly indicates that it costs about Rs. 1.26 per bottle of juice (average of April 1993 to August 1993) and at the rate of Rs. 2 per bottle (as it is being priced), the society earns a profit of 74 paise per bottle i.e. 37 per cent. This is to suggest that it is profitable to take up processing of fruits on a large scale and concurrent improvements need to be made for the marketing of the processed products.

Table 10.7: Economics of Juice Preparation

Sl. No.	*Items of expenditure*	*Amount (Rs.)*
1.	Sugar	20359.60
2.	Grapes	27771.20
3.	Establishment	6557.12
4.	Others	44.324.60
	Total	74012.52
No. of bottles prepared		58604
Unit cost (Rs/bottle)		1.26*
Profit (Rs/bottle)		0.74
@ Rs. 2/bottle		(37%)

*Includes straws, tonovin, caps, colour, firewood, ice etc.

Financial Performance of HOPCOMS

Financial Transactions of HOPCOMS

The society has made good progress in the recent years, except in the year 1991-92 when it incurred heavy losses. A comparison of the progress of the society in the 70s and at present is made in Table 10.8 in terms of net profit/losses of the society. It could be observed that HOPCOMS (then BGGCOMS) has moved from the position of a perpetual loser in the 70s (*Subrahmanyam et al., 1979*) to a position of profit maker. The main reason for this has been the positive change in the policies of HOPCOMS with respect to procurement and sales, higher profit margin and increased turnover. In the 70s, it was observed that the society incurred losses due to the wrong policies followed by it, poor recovery of the loans from the members and also due to the disproportionately high overheads (*Subrahmanyam et al., 1979*). It seems, HOPCOMS has acquired the business acumen over the years and started the sale of fruits and vegetables through a wider network of retail outlets and credit sales to bulk buyers. It is interesting, however, to note that the society incurred heavy losses in 1991-92 despite its Mysore branch made profits. Thus, to probe deep into the factors responsible for the significant changes in the performance of the society, an analysis of the trading account is also made in Table 10.8.

Table 10.8 : Financial Performance of HOPCOMS

Year	*Turnover (Rs. lakh)*	*Trade profit (Rs.)*		*Overheads (Rs.)*		*Net profit (Rs.)*
1970-71	17.82	16378	(9.19)	229904	(12.90)	–25794
1975-76	47.00	313214	(6.66)	436013	(9.28)	-85841
1987-88	770.95	8152654	(10.57)	7382481	(9.58)	931112
1988-89	1055.50	9498348	(9.00)	8894709	(8.43)	750218
1989-90	941.06	9451782	(10.04)	8918409	(9.48)	945159
1990-91	1418.05	14014886	(9.88)	13429142	(9.47)	876997
1991-92	1763.92	17675749	(10.02)	22979699	(13.33)	–2837819
1992-93	2104.93	24137171	(11.47)	24634972	(11.70)	118882, –(2858983)*

Source : 1. Subrahmanyam *et al.*, (1979) for 1970-71 and 1975-76.

2. HOPCOMS, Bangalore, annual reports for 1987-88 to 1992-93.

Note : Figures in parentheses are percentages to turnover.

*Cumulative loss as on 31-3-1993.

It may be observed from the table that HOPCOMS has increased its turnover considerably. A corresponding increase was also observed both in trade profits and overheads. An analysis of influence of trade profits and overheads on net profit of the society made in Table 10.8 reveals the following:

(i) Whenever the proportion of the overheads exceeds that of the trade profit, then the society incurs loss, as has been the case in 1970-71, 1975-76 and 1991-92. This, perhaps, is due to the fact that trading is the main activity and trade profit is the main source of receipts of the society. Trade profit forms over 90 per cent of the receipt of the society and in order to enable the society to make net profits, the overheads have to be proportionately low as has been the case in the years 1987-88 to 1990-91.

(ii) Whenever the rate of profit is significantly low, it results in net losses, though the overhead expenses are low. For example, in 1975-76, the overheads have decreased by 3.62 per cent from 12.9 per cent in 1970-71. But, the trade profit declined by about 2.53 per cent. The net loss in 1975-76 was, however, substantially high compared to 1970-71. Thus, it may be inferred that if the trade profit forms a small fraction of the turnover (as in 1975-76), it results in net losses to the society despite decrease in the overheads.

The branchwise details presented in Table 10.9 also indicate that while all other branches incurred heavy losses, the Mysore branch made profits as the overheads were kept proportionately low. A detailed analysis of two branches, viz. Bangalore, which forms about 80 per cent of the turnover and Mangalore, which accounts for only 7 per cent of the turnover, was made to further strengthen our arguments. Our earlier finding that overheads in excess of the trade profits leads to net losses to the society is vindicated from the illustrations.

A further analysis of the components of the overheads revealed that the establishment cost has increased in 1991-92 as majority of the employees were confirmed and thus became eligible for provident fund and other benefits. As a matter of fact, the salary of the staff has increased from Rs. 62.78 lakh in 1990-91 to Rs. 121.15 lakh in 1991-92 and depreciation on fixed assets increased from Rs. 8.32 lakh to Rs. 21.69 lakh during the said period. Besides, a fire accident at Vijayapur godown and the repair cost of the cold storage in Bangalore also added to the contingency of the society. However, considering the phenomenal increase in the turnover (from Rs. 14 crore to Rs. 17.6 crore) and also increase in the margin from 20 per cent to 25 per cent, we feel that, had the society been 'managed' well, the trade profits could have offset the additional overheads and this would have left the society with net profits.

Table 10.9 : Branchwise Performance of HOPCOMS (1991-92)

(Rs. lakh)

Branch	*Turnover*	*Trade profit*	*Overheads*	*Net profit*
Bangalore	1384.28	133.52 (9.65)	181.86 (13.14)	–31.58
Mysore	208.34	31.45 (15.10)	30.15 (14.47)	4.56
Mangalore	99.73	8.17 (8.19)	12.14 (12.17)	–3.13
Kolar	42.17	4.21 (9.98)	7.31 (14.84)	–1.15
Mandya	29.41	–0.60	3.59 (12.21)	–3.24
Total	1763.93	176.75 (10.02)	235.05 (13.33)	–34.54

Note : Figures in parentheses are percentages to total turnover (–) loss.

Ratio Analysis

In order to understand the financial soundness of HOPCOMS, certain accountancy ratios were worked out from the balance sheet of the society and these ratios are presented in Table 10.10. A brief interpretation of them is given below :

(i) *Current Ratio* : Also called as working capital ratio, this is obtained as the ratio of current assets to current liabilities. As is evident from the table, the ideal ratio of 2 : 1 has never been maintained. However, 1989 onwards, the ratio is higher than 1 thereby indicating that, of late, the society has been able to meet its current obligations and to expand its business gradually.

(ii) *Quick Ratio* : This is also called as liquidity ratio and is calculated as the ratio of liquid assets (current assets-stock) to current liabilities. It is clear from the table that there has been an improvement in the liquidity of the society as the ratio is higher than the ideal ratio of 1 : 1 in the recent years.

(iii) *Total Assets to Total Liabilities* : It may be observed from the table that the ratio has been just above 1 up to 1990-91

Table 10.10 : Accountancy Ratios of HOPCOMS

Ratio	*1970-71*	*1975-76*	*1987-88*	*1988-89*	*1989-90*	*1990-91*	*1991-92*	*1992-93*
1. Current Ratio	0.9800	0.9200	1.1400	0.7241	1.2600	1.1600	1.1400	1.7594
2. Quick Ratio	0.9200	0.8200	0.8922	0.4404	1.0445	0.8539	1.0029	1.4627
3. Total assets to total liabilities	1.0300	1.0800	1.0693	1.0893	1.1160	1.1230	0.9567	0.9609
4. Total liabilities to owned fund	18.4700	14.4700	1.7788	2.0610	1.8012	2.0477	2.3366	2.0670
5. Current liabilities to owned fund	5.9500	5.9100	0.5553	0.7576	0.7142	0.6444	0.8859	0.3487
6. Fixed assets to owned fund	0.5100	1.3500	0.7130	0.7843	0.6861	0.7804	0.7240	0.8336
7. Fixed assets to total assets	0.6595	0.5876	0.3749	0.3592	0.3412	0.3394	0.3239	0.4197
8. Net profit to total sales	—	—	0.0121	0.0071	0.0100	0.0062	—	0.0006
9. Net profit to owned fund	—	—	0.1159	0.0776	0.0737	0.0619	—	0.0034

and is less than 1 since 1991-92, which indicates that without the huge capital contribution from the Government, it would not have been possible for the society to expand its business and make profits over the years.

(iv) *Total Liabilities to Owned Fund* : This ratio was worked out to know as to whether or not the society has been able to meet its commitments from the owned fund (share capital), reserves etc.,) itself. It may be seen that in the 70s, the society was able to meet only 1/15th or 1/18th of its commitments from the owned fund. However, in the 80s as the society consolidated its position, there has been a considerable decline in the ratio to around 2 : 1 which, thus, suggests that the society is now able to meet around 50 per cent of its commitments from the owned fund. This also suggests that the owned fund of the society is still not sufficient. Hence, there is a need to increase the share capital of the society, preferably from the producer members.

(v) *Fixed Assets to Owned Fund* : The ratio was worked out to know the extent of investment of owned fund in fixed assets. As can be seen from the table over 80 per cent of the owned fund is invested in fixed assets. However, as may be seen from the table, the proportion of the fixed assets to total assets has come down from 60 per cent in the 70s to about 40 per cent in the 80s and 90s. Thus, it suggests that about 60 per cent of the total assets is in semi-liquid or liquid form. However, over investment of owned fund in fixed assets is not desirable as it reduces the availability of working capital to the society.

MAJOR FINDINGS

1. For a state level cooperative society like HOPCOMS, the membership of 6200 is too meagre. Efforts should be made to increase the membership. This can be achieved by making it mandatory on the part of the non-member producers supplying fruits and vegetables to the society to become the members. It should be made as a policy to accept only from the member producers.
2. The society should enroll more members, especially in the districts where it has branches, and hence raise capital. This

will reduce the dependence of the society on the Government.

3. Acceptance of fruits and vegetables only from the members will help the society to regulate the supply. Depending on the requirements, the society can also suggest vegetable crop plans to its members.
4. The society should accept all the produce brought by the producer member and pricing should be based on the quality of the produce.
5. As the society has incurred heavy losses in 1991-92 which could not be wiped out even in 1992-93 due to heavy overhead expenses, there is a need to increase its sales and hence the trade profits. Sales can be increased by having contractual arrangements with the processors.
6. The society can also think of packed sales of fruits and vegetables taking into account the additional costs.

11

CASE STUDY 2

THE DAKSHINA KANNADA DISTRICT HORTICULTURE PRODUCE GROWERS COOPERATIVE MARKETING AND PROCESSING SOCIETY LTD. (DKHOPGCOMS), MANGALORE, KARNATAKA

Realising the importance of cooperatives in fruit and vegetable marketing, in the state of Karnataka, district societies were organised on cooperative lines in the mid sixties for marketing of fruits and vegetables grown in the respective districts. However, despite being in existence for the past 30 years, the performance of these societies has been very unsatisfactory. The Dakshina Kannada District Horticulture Produce Growers' Cooperative Marketing and Processing Society (DKHOPGCOMS) is one such society which has been incurring losses ever since its establishment in 1965. Meanwhile a branch of the Horticultural Producers Cooperative Marketing and Processing Society (HOPCOMS), which came into being in 1987, has been performing well though both the societies were established for the same purpose of marketing of fruits and vegetables. Hence it is envisaged to study in detail the performance of DKHOPGCOMS *vis-a-vis* HOPCOMS (Mangalore) in terms of policies followed, factors responsible for the differed performance etc., by analysing the data on organisational, business and financial aspects.

Background

The present DKHOPGCOMS was established on 24-06-1965 with the main objective of arranging for the sale of horticulture produce of the members; advancing short term loans to the members; supplying the production requisites and to provide technical know-how on modern practices. The society was started by the Department of Horticulture at Mangalore (Dakshina Kannada District) with Deputy Director of Horticulture as the Ex-Officio president of the society. However, since 1993 the president is elected from among the 'A' class members. Although the jurisdiction of the society extends to the whole of Dakshina Kannada district, presently, the society operates only in Mangalore, Udupi and Kundapur taluks.

Membership and Share Capital

At the time of establishment, the society was having 40 members with a paid-up capital of Rs. 2180 which increased to 1040 and Rs. 83,475 respectively in 1987-88. Between 1987-88 and 1992-93, the membership and the paid-up share capital have increased to 1547 and Rs. 96,675 respectively. It was observed that Government contributed about 52 per cent towards the share capital of the society and 'A' class membership has accounted for about 43 per cent.

Business Performance

Procurement of Fruits and Vegetables

The society purchases fruits and vegetables both from the members and the market. The produce from the members is procured through two procurement centres located at Mangalore and Kundapur. It was observed that about 50 per cent of the produce is procured from the market as against about 30 per cent in case of HOPCOMS (Mangalore). It was reported (Subrahmanyam *et al.*, 1994) that market purchases result in additional cost of Rs. 1-2 per kg in case of vegetables and Rs. 3-5 per kg in case of fruits. Further, DKHOPGCOMS does not charge for the transport of fruits and vegetables from the members whereas HOPCOMS (Mangalore) deducts the actual cost of transport from the sale proceeds while making payment to the growers. The policy of not charging the transport cost was found to be one of the reasons for the poor performance of FRESH, Hyderabad (*Sudha* and *Subrahmanyam, 1993*). Hence, it is desirable that DKHOPGCOMS purchases its produce mostly from the field rather than from the market and while procuring from the field transport cost is charged.

Sales of Fruits and Vegetables

Fruits and vegetables purchased by the society are disposed of through: (1) retail outlets on cash basis, and (2) bulk sale to hostels on credit basis. It may be noted from Table 11.1 that upto 60-85 per cent of the sales of DKHOPGCOMS was on credit basis and the same in case of HOPCOMS (Mangalore) was at maximum of 21 per cent. The emphasis on credit sales in case of DKHOPGCOMS resulted in less availability of capital and this prevented the society from increasing its procurement of fruits and vegetables which is hardly 1-1.5 tonnes per day as against around 3 tonnes per day in case of HOPCOMS (Mangalore). This suggests that there is a need to increase the cash sales of fruits and vegetables by DKHOPGCOMS so as to improve the overall performance of the society. Further, we noted that DKHOPGCOMS has only 6 retail outlets as against 20 in case HOPCOMS (Mangalore). We also observed that HOPCOMS (Mangalore) has two outlets in Udupi where DKHOPGCOMS wants to open its outlets. In Manipal, HOPCOMS (Mangalore) plans to open its outlets which would be overlapping with the already existing outlet of DKHOPGCOMS. Lack of coordination appears to be the cause for this type of dichotomy. Instead, we feel that the societies should open their outlets in such a way that it will not overlap the existing one. This would help in proper distribution of business and also avoids the unhealthy competition between them.

Table 11.1 : Proportion of Sales through different Modes

(Per cent)

Year	*Retail outlet (Cash)*		*Hostels (Credit)*	
	DKHOPGCOMS	*HOPCOMS (Mangalore)*	*DKHOPGCOMS*	*HOPCOMS (Mangalore)*
1987-88	40	—	60	—
1988-89	30	—	70	—
1989-90	25	82	75	18
1990-91	15	79	85	21
1991-92	23	82	77	18
1992-93	34	84	66	16

Price Policy

Both DKHOPGCOMS and HOPCOMS (Mangalore) pay the prevailing wholesale price to the growers. Analysis of price data indicated that HOPCOMS (Mangalore) adds 25 per cent (33 per cent in case of tomato) to the sum total of procurement price + transport cost to arrive at the retail price. On the other hand it was observed that DKHOPGCOMS fixes its retail price arbitrarily though the society claims that it retains a margin of 20 per cent while fixing the retail price.

Diversification of Business

The district society has diversified its business since 1990-91 in favour of production related activities. It started distributing grafted plants of sapota and mango to its members. In 1992-93 it also supplied vegetable seeds, fertilisers, plant protection chemicals and mini kits prepared by the Department of Horticulture (Table 11.2). It may be noted from the table that by diversifying the business though the percentage of fruits and vegetables sold decreased (from 89 to 39) there was a commendable increase in the total sales of the society. This much needed increase in the sales also resulted in positive profit of Rs. 23,577 in 1992-93.

Table 11.2 : Progress of Sales of DKHOPGCOMS

(Rs.)

Year	*Fruits and Vegetables*	*Grafted plants*	*Minikits*	*Fertilisers and PPC*	*Vegetable Seeds*	*Total sales*
1987-88	260252 (100)	—	—	—	—	260252
1988-89	284262 (100)	—	—	—	—	284262
1989-90	224854 (100)	—	—	—	—	224854
1990-91	349597 (89)	44400 (11)	—	—	—	393997
1991-92	322668 (85)	55240 (15)	—	—	—	377908
1992-93	555675 (39)	309708 (21.5)	21361 (1)	537268 (38)	4150 (0.5)	1428162

*Figures in parentheses are percentages to total sales.

Financial Performance

The financial performance was analysed in terms of growth in turnover, proportion of overheads and trade profits in turnover, net profits over the years etc. It may be noted from Table 11.3 that both the societies have increased their turnover between 1987-88 and 1992-93. However, it was observed that the turnover of DKHOPGCOMS was considerably

Table 11.3 : Financial Performance of the Societies

Year	DKHOPGCOMS				HOPCOMS (Mangalore)			
	Turnover (Rs)	*Trade profit as % of turnover*	*Overheads as % of turnover*	*Net profit (Rs.)*	*Turnover (Rs.)*	*Trade profit as % of turnover*	*Overheads as % of turnover*	*Net profit (Rs.)*
1987-88	263561	10.23	33.35	–57659	3188279	8.78	6.51	77982
1988-89	288849	12.60	28.66	–41734	7313147	6.13	6.13	7485
1989-90	225611	11.40	28.93	–38597	6713436	10.21	8.38	138062
1990-91	395809	13.48	17.55	–13484	9329541	10.09	9.86	34810
1991-92	381038	11.46	21.49	–34568	9972514	8.19	12.17	–312846
1992-93	1460356	9.62	8.20	23577	13693228	12.48	12.26	40470

(–) = loss

low as compared to that of HOPCOMS (Mangalore). While HOPCOMS (Mangalore) could make profits throughout except in 1991-92, DKHOPGCOMS incurred losses throughout except in 1992-93. The analysis of the overheads and the trade profits would reveal that despite making trade profits DKHOPGCOMS incurred losses as it was not sufficient to offset the proportionately high overheads. On the other hand HOPCOMS (Mangalore) could make profits as it could maintain the overheads proportionately low in relation to the trade profits. Thus, in order for DKHOPGCOMS to make profits there is a need to keep the overheads comparatively low in relation to the trade profits.

Conclusions

1. The diversification of business towards production requisites has improved the turnover of DKHOPGCOMS. This also helped the growers to get the inputs at a reasonable price.
2. The free transport of fruits and vegetables of the members should be discontinued and checked.
3. The society's credit sales is responsible for the blocking in of the capital which could be used for enhancing the business of the society.
4. The overheads were found to be higher compared to the trade profits.
5. The society's lack of coordination with HOPCOMS (Mangalore) resulted in duplication of retailing business.

12

CASE STUDY 3

THE FARMERS' RURAL EXTENSION SERVICE IN HORTICULTURE MARKETING COOPERATIVE SOCIETY LTD. (FRESH), HYDERABAD, ANDHRA PRADESH

The Farmers' Rural Extension Service in Horticulture (FRESH) marketing society (Hyderabad) has been chosen purposively, as it is one of the new cooperative fruit and vegetable marketing societies which has been started in Hyderabad (AP) recently with the financial assistance from NCDC on the lines of HOPCOMS after seeing its success. But even after five years of establishment it has not made significant profits and is yet to break even. Hence a detailed study of this society was taken up.

Organisational Aspects of "FRESH" (Hyderabad)

The Farmers Rural Extension Service in Horticulture (FRESH) marketing co-operative society, Hyderabad, registered under the AP Co-operative Societies Act, in July 1986, became functional from February, 1987.

Membership and Share Capital

The pilot project sponsored by State Government and financed by NCDC, began with a paid-up share capital of Rs. 2.25 lakhs from the Government. The initial membership of 34 producers and consumers by the end of 1986, grew up to 668 by February 1992, 663 of this constituting "A" Class Members of growers, consumers and traders, 4 co-operative societies as "B" Class members apart from the State Horticultural Department.

The paid-up share capital of the society as on 30-6-1991 is Rs. 10.03 lakh, with Rs. 5.66 lakh from sale of "A" Class shares worth Rs. 100 each, to producers and consumers, Rs. 40,000 from co-operative societies and Rs. 3.97 lakh from State Government (Table 12.1). Besides this, the society also has received financial assistance from several organisations like NCDC, Social Welfare Department, Co-operative Department, National Horticultural Board, etc. to the tune of Rs. 126.056 lakh (30-6-90), either in the form of share capital, subsidy, margin money or long term loan. The society has an outstanding loan of Rs. 57.80 lakh.

Table 12.1 : Membership and Share Capital Position of FRESH

S. No.	*Years*	*Members enrolled*	*Share capital (Rs.)*	*Govt. share*	*Govt. share as % of total*
1.	1986-87	34	225000	225000	100.00
2.	1987-88	252	278050	225000	80.92
3.	1988-89	266	407890	225000	55.16
4.	1989-90	668	916900	311000	33.92
5.	1990-91	668	1003000	397000	39.58
6.	1991-92	668	1003000	397000	39.58

Jurisdiction and Objectives

As of now, the society is operational only in the twin cities of Hyderabad and Secunderabad in Andhra Pradesh State. The prime objective of the society is to assist the producer in getting a better share of consumer's rupee and the consumer in getting a quality produce worth its price. The society also aims at improving the crop productivity of its member growers in the long run through input supply and imparting appropriate technical guidance.

Management

The society has an executive committee of eleven members, 4 elected from growers, 4 consumers and 3 nominees of the Government Department. The Chairman of the society is an elected member among these, while the Managing Director and the Accounts Officer are the ex-officio members from the State Department on deputation. The society has 110 paid employees for running the day-to-day business.

Business Aspects of the Society

The channels followed for procurement of fruits and vegetables and the mode of disposal adopted by FRESH is presented in Fig. 12.1 and the same is discussed briefly below:

Procurement of Fruits and Vegetables

The society procures fruits and vegetables from member growers/non-members (a) by field procurement, (b) by participating in the open market auction and (c) from societies of other states.

Field Procurement : The society has three procurement centres at the villages viz., Devarayamjal, Godhimakunta and Madannapet, within a radius of 30 km from the twin cities, and caters to the needs of a total of 38 villages located within a distance of 9 km from each of these procurement centres. The society collects all the produce brought to these centres, from its member growers and at times even from non-members, depending largely on the type of fruit or vegetable. The spoiled and rotten lots are rejected, while the rest get paid based on the previous day's wholesale market price. The produce then gets transported to the society's processing centre (as it is called) at Moulali, by society's own vehicles or by hired trucks (whose transportation costs are borne by the society). The produce washed, sorted and graded at the factory is then sold through the society's retail outlets.

Procurement from Other Sources: The society also participates in the wholesale market auctions, in order to procure some fruits and vegetables which are not available at field, in order to meet the prevailing demand from some bulk consumers. Apart from this, the society of late has also started procuring from organisations of other states like the HPMC or HOPCOMS as well.

Sales of Fruits and Vegetables

The society undertakes its marketing operations through 31 retail outlets

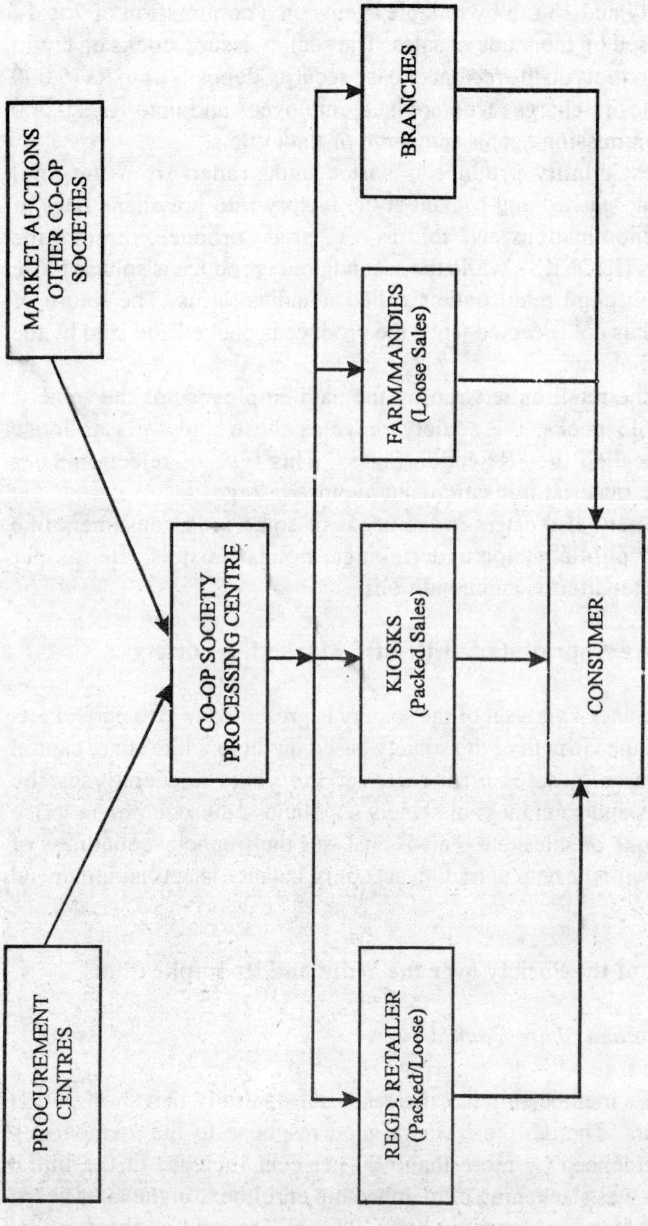

Fig. 12.1 : Marketing Channels of FRESH

(KIOSKS, farms and mandies), 13 of which managed by paid employees of the society and 18 run by reliable agents on a commission of 7 to 14 per cent, based on the mode of sales. The society issues stocks on credit to the retail outlets on the receipt of cash security deposits upto Rs. 5,000 from the sales in-charges who are paid employees and upto Rs. 20,000 from the commission agents in-charge of the outlets.

The best quality produce is sorted and graded from the total procurement, washed and packed at the factory into polythene bags of different denominations and sold as "A" grade produce, through the retail outlets (KIOSKS), while the second grade produce is sold as loose vegetables through retail outlets called mandies/farms. The return of unsold stock is not accepted when the produce is packed and sold by the commission agents.

When the retail outlets run by the paid employees of the society, return unsold packs, the society recycles these and sells as loose vegetables called the "Rejection Sales". This type of rejection sales improve the material utilisation and avoids wastage.

The society also caters to the needs of some bulk consumers like canteens of public sector undertakings, hotels, hostels, etc. as per mutually agreed terms and conditions.

Performance Appraisal of "FRESH" Marketing Society

The performance appraisal of the society is presented in two parts. Part-I deals with the Growth of the society based on factors like share capital and membership, sales turnover over the years and analyses the implications and soundness of society's policies on procurement, price fixation, mode of sales, etc. Part-II analyses the financial soundness of the society with the help of trading accounts, balance sheets and financial ratios.

A. Growth of the Society over the Years and its Implications

Membership and Share Capital

The society's membership has increased substantially in a short period of four years. Though there was a good response to the membership drive as evidenced by more than 600 per cent increase in the initial years, there was slackening of membership enrolment in the later years, as is evident from the stagnation after 1989-90. The paid-up share capital of the society has also increased from Rs. 2.25 lakh in 1986 to Rs. 10.03 lakh by 1991. The Government share in the share capital in 1986, was Rs. 2.25 lakh, and it has increased to only 3.97 lakh by 1991, accounting for only 39.58 per cent of total share capital. Though, this decreasing Government share indicates fairly good growth in membership and share capital position of the society, it was not sustained during the last two years, due to stagnation of membership (Table 12.1).

Field Procurement

Table 12.2 presents the details of procurement of fruits and vegetables by the society during the co-operative year, 1988-89, 1989-90 and 1990-91. It could be observed that the society's total procurement in the three years from 1988-89 to 1990-91 has increased from 22 lakh kg to 32 lakh kg. However, the procurement from field has shown a decreasing trend over the years. Except in the first year, it has never exceeded 50 per cent of the total procurement in any month, and it was as low as 26 per cent during January 1990. This clearly shows that the society is depending more on market purchases rather than procurement from cultivators/growers, thus, showing that the main objective of helping the cultivators could not be fulfilled and this may be one of the reasons for the society not being able to attract the cultivators as evidenced by the stagnation of membership and offer itself as a better alternate channel of marketing. The studies in the past have also shown that such increased dependency on the market procurement rather than from cultivators is not desirable and this was found to be one of the reasons for the society running into losses (*Subrahmanyam et al., 1979*).

Table 12.2 : Share of Field Procurement in Total Procurement

S. No.	*Co-operative year (July-June)*	*Total Quantity Procured (lakh/kg)*	*Per cent share of Field procurement*
1.	1988-89	22.63	51.93 (28-100)*
2.	1989-90	24.11	35.38 (26-48)
3.	1990-91	31.41	39.48 (33-50)

*Figures in parentheses show the range.

Price Fixation Policy

The society offers previous day's wholesale market price to the cultivators at farm gate or the village, then adds 40-60 per cent to the procurement price and fixes the retail price in such a way to get atleast 15-20 per cent margin after meeting all the trading expenses and wastage.

By this procurement price policy, the society incurs heavy expenditure in transporting the produce from village level procurement centres to the factory before the produce is sent to retail outlets. Since, it is the commission charges that account for 40-50 per cent of the total marketing costs (*Subrahmanyam, 1981*) the society is reducing a great burden of commission, loading and unloading charges for the cultivators by offering wholesale price at the village. Hence, instead of bearing the entire burden of transportation costs, the society could work out unit cost (Rs/km/kg) for all the routes and collect a portion of it from the cultivators. By doing so, the cultivators will still be getting at least 40 per cent more than the open market price. Encouraging the cultivators to supply straight to the factory; as it is done by other co-operatives, is beneficial for the society in the long run and will help in covering more number of cultivators from other areas as well.

Mode of Sales and its Impact

The details of sales through different modes (KIOSKS or Mandies), and the price realised is presented in Table 12.3. The society's sale of fruits and vegetables has increased from 17.91 lakh kg during the co-operative year 1989-90 to 25 lakh kg in 1990-91 and 19 lakh kg just in 9 months during 1991-92. Regarding sales through different modes, it could be observed that loose sales through retail outlets called farms or mandies account for 58-65 per cent of the total sales, followed by packed sales through KIOSKS (16-25%). Over the years, it could be seen that the loose sales/day and rejection sales have shown marked growth, while sales through KIOSKS have shown a declining trend. As against an increase of 49.83 per cent in the loose sales, the per day sales through KIOSKS declined by 10.48 per cent during 1989-90 to 1990-91. Even in the month of March 1992, when the total sales/day have increased to 10.80 t/day, the loose sales and rejection sales have increased to 4.62 and 3.74 t/day, while the packed sales is only 1.02 t/day. This clearly shows that sales in packed form is not encouraging and it is not favoured by the consumers.

Table 12.3 : Share of Sales through Different Modes and the Prices realised

Sl. No.	*Year (July-June)*	*Total*			*Sale through Different Modes and Sale Price*					
		Sales (lakh kg)	*Loose*		*Packed*		*Bulk*		*Rejection*	
			% Sale	*Sale price*	*% Sale*	*Sale price*	*% Sale*	*Sale price*	*% Sale*	*Sale price*
	1	2*	3*	4**	5*	6**	7*	8**	9*	10**
1.	1989-90	17.92	60	4.72	25	4.92	6	3.58	9	0.76
		(4.98)	(3.01)	(2.83)	(1.24)	(1.22)	(0.28)	(0.20)	(0.44)	(0.07)
2.	1990-91	24.98	65	4.66	16	5.34	5	4.36	10	1.39
		(6.94)	(4.51)	(3.03)	(1.11)	(0.85)	(0.33)	(0.22)	(0.74)	(0.14)
3.	1991-92	19.09	58	5.43	19	5.81	5	4.66	16	1.76
		(7.07)	(4.08)	(3.14)	(1.33)	(1.09)	(0.39)	(0.25)	(1.16)	(0.28)

*Figures in parentheses in columns 2, 3, 5, 7 & 9 show the Average daily sales through the respective modes (t/day).

**Figures in parentheses in columns 4, 6, 8 & 10 show the weighted price realised (Rs/kg).

Volume of Sales and its Impact : The relative profitability of different modes of sale was estimated by calculating weighted price realised in different modes of sales during 1989-90. The weighted price was calculated by the following method:

$$P_{wt} = P_i \times Q_i$$

P_{wt} = Weighted price
P = Selling price
Q = Percentage quantity sold
i = 1-3, different modes.

The results are presented in Table 12.3.

It could be seen from Table 12.3, that though the unit price is the highest, for example Rs. 4.92 during 1989-90, in case of KIOSK sales, the weighted price/unit realised is only Rs. 1.22 as only 25 per cent of the total quantity is sold in packed form. The value addition/kg, on the other hand is highest when sold as loose vegetables through mandies/ farms, ranging from Rs. 2.51 to 3.37/kg during 1989-90, or Rs. 3.03 and Rs. 3.14 in 1990-91 and 1991-92. Though the absolute price difference between packed sales and loose sales is not more than Rs. 0.60/kg, the loss in value realised per kg, by virtue of lower per cent sales in packed form is even upto Rs. 2.00 compared to the loose sales.

Economics of Sales through Different Modes : The additional costs and returns involved in packed sales as compared to loose sales is calculated and presented in Table 12.4. It could be seen from the table that the society is actually losing nearly 20 paise/kg in packed sales as compared to loose sales, showing that the society has not properly taken into account the packing costs like labour, depreciation of packaging machinery etc., while fixing the sale price through KIOSKS.

Table 12.4 : Economics of Packing *Vs.* Loose Sales (1989-90)

(*Rs./kg.*)

S. No.	*Mode of sale*	*Costs*			*Returns*	*Net Gain*
		Procure-ment price	*Packing material cost*	*Total cost*	*Sale price*	*Difference*
		(1)	(2)	3 (1+2)	(4)	5 (4– 3)
1.	Packed Sales (KIOSKS)	2.51	0.40	2.90	4.92	2.02
2.	Loose Sales (FARMS)	2.51	—	2.51	4.72	2.21

*Based on the information given in the Society's reports.

Based on the low volume of sales, as well as loss in the returns to the society, it is clear that selling in packed form is not desirable both from the view point of society as well as consumers.

Financial Performance of the Society

From the trading details presented in Table 12.5, it could be observed that in the alternate years, the society has earned net profit from the sale transactions. Though the price policy adopted by the society is supposed to give 15-20 per cent margin after meeting all the expenditure connected with sales, the society could get positive margin of 15 per cent only in 1990. Even during this year, on the whole, the society had made a loss when the expenditure towards establishment and contingency etc., were accounted. During the four years of operation, the society was making cumulative losses to the tune of Rs. 12 lakh and the yearly profit earned during alternate years could not offset the previous years losses. From the details of the accounts it is clear that the society needs better performance in its trade transactions to meet the high overhead costs.

Table 12.5 : Trading Account of "FRESH" Marketing Society (Hyderabad)

S. No.	*Items*	*Years*			
		1987	*1988*	*1989*	*1990*
A.	Total Income	945749	4820943	5833762	8641412
	i. Sales	907636	4740563	5798507	8594348
	ii. Closing stock	38113	80380	35255	47064
B.	Expenditure	908978	4361775	5935294	6922477
	i. Purchases	908978	4323662	5854914	6887222
	ii. Opening Balance	—	38113	80380	35255
C.	Trading Overheads	147592	67828	185159	592422
D.	Total Expenditure (B+C)	1056570	4429603	6120453	7514899
E.	Trade Profit/Loss (A-D)	-110821	391340	-286691	1126513
F.	% Margin Earned (E/D)	-10.49	8.83	-4.68	14.99
G.	Other Income	3471	118642	357319	264994
H.	Establishment & Contingency	368784	87605	207061	1987377
I.	Gross Profit/Loss (E+G-H)	-476134	422377	-136433	-595870
J.	Interest & Depreciation	400905	624050	0	120555
K.	Net Profit/Loss (I-J)	-877039	-201673	-136433	-716425
L.	Reserve	193363	24326	31986	28787
M.	Carry Over P/L	0	-1070402	-296401	-464820
N.	Grants-in-aid	0	1000000	0	0
O.	Cumm. Adjusted P/L	-1070402	-296401	-464820	-1210032

*Except for the item F, which was calculated, all other informations are as per the Society's records.

Financial Ratio Analysis

To evaluate the financial standing of the society, some of the ratios which will be useful for appraisal like current assets to liabilities ratio, equity ratio, etc., were worked out and presented in Table 12.6.

Table 12.6 : Financial Standing of FRESH

S. No.	Ratio	Years			
		1987	*1988*	*1989*	*1990*
1.	Current assets to current Liabilities	1.28	—	0.06	0.30
2.	Current Liability to Tangible net worth	0.04	—	0.57	0.30
3.	Current assets to total assets	0.05	0.06	0.01	0.02
4.	Fixed assets to owned funds	0.27	0.25	0.76	0.80
5.	Total Liabilities to owned funds	1.52	1.89	2.86	4.08
6.	Total assets to total Liabilities	0.65	0.96	0.94	0.89

The fact that both the ratio of the current assets to current liabilities and current liabilities to tangible net worth, is less than one, shows that the society will not be able to meet its liabilities both in short run and long run. The increase of total liabilities to own fund ratio over the years shows that the stake of the borrowers is more and that of society's is less, which will burden the society in terms of payment of huge interest thus, eating away its profits which is not desirable. So, on the whole based on these ratios it can be concluded that at present the financial standing of the society is not encouraging.

Conclusions

From the detailed analysis of FRESH society presented above, the following conclusions are drawn:

1. It is evident from the fact that even though, the society had made profit from the trade transactions in some years, still overall, the society has incurred net losses due to the huge overheads.
2. Because of the huge borrowings and blocking in of capital assets which involved the fixed commitments like interest and depreciation, has offset the profits earned from trade transactions.

3. The society's policies like not charging transport costs during procurement and fixing prices for the packed vegetables without considering additional costs and returns, seem to be responsible for the society not making desirable profit margins.
4. The indulgence in large scale market purchases and less dependence on direct procurement from the cultivators also contributed to the losses of the society.
5. The fact that the packed sales done by the FRESH cooperative society has declined over the years whereas the loose sales have increased, shows that the consumers of the twin cities did not prefer the former. This could be attributed to the consumers' psychology that they were denied the satisfaction of choosing the produce when packed and sold in polythene packs.

13

CASE STUDY 4

COOPERATIVE FRUIT (BANANA) SALE SOCIETIES JALGAON, MAHARASHTRA

Banana is one of the most nutritious and the oldest cultivated tropical fruits in India. Jalgaon district in Maharashtra has been specialised in the commercial cultivation of 'Basrai' (dwarf cavendish) banana which is produced throughout the year and most of the northern markets get their supplies from here.

The marketing of Jalgaon banana is done through three different agencies viz. Cooperative Fruit Sale Societies, the Group Sale Agencies and the Private Traders. The cooperatives are established by the banana growers with the objectives of taking up marketing of banana in different markets and supply of production requisites like fertilisers and credit facilities to the grower-members. The group sale agency is formed by two or three landlords and/or merchants and the private agency involves the local persons popularly known as *Bagwans'* [see *Waskar* and *Roy (1993)* and *Biradar* and *Kasar (1983)*].

While bulk of the Jalgaon banana is marketed in Delhi (New Azadpur market) through commission agents of the respective marketing agency, a small quantity is also disposed of in the local market itself. The studies on marketing of Jalgaon banana have indicated that the cooperatives performed slightly better compared to the other two agencies in terms of relatively low marketing cost, higher producer's share in consumer rupee and slightly higher price for banana [*see Biradar* and *Kasar (1983), Nawadkar et al., (1990), Tilekar* and *Pawar (1989), Pawar et al., (1989)* and *Pawar (1991)*]. The studies also revealed that though cooperatives could get better prices to their members compared to their counterparts, the policies followed differed from one society to another resulting in differences in prices paid to their members. Hence, it was felt appropriate to study in detail the policies followed by a few cooperative fruit sale societies operating in the area.

Data and Methodology

Two cooperative societies dealing in banana, namely, The Kissan Cooperative Fruit Sale Society Ltd. (KCFSSL), Nimbhora located in Raver Taluk, established in 1972 and the Jalgaon Fruit and Vegetable Sale Society Ltd. (JCFVSSL) located in Jalgaon and established way back in 1939 were purposively selected from among around 35 cooperative fruit sale societies dealing in banana in Jalgaon district. These two societies have different marketing policies in disposal of banana fruits. The Kissan society is marketing most of the produce in Delhi market whereas the Jalgaon society is concentrating in the local market. Hence, detailed information regarding the functioning of the society mainly the membership, grading and price policy followed, financial performance etc., were collected from the records of these societies for a period of 11 years from 1982-83 to 1992-93. The data was analysed in terms of growth rates, accountancy ratios etc.,. The turn over figures were deflated by the respective indicators of prices of food articles (which also includes banana) to account for inflation so as to arrive at the compound growth rates in real terms.

Results and Discussion

Share Value and Membership

The membership of both the societies consists of permanent members and the Government. Both the societies have the practice of enrolling those who sell banana occasionally through them as nominal members who do not have the voting rights and are not entitled to dividend or the facilities like supply of fertiliser and advance which the permanent members enjoy. In 1992-93, there were 83 nominal members in Kissan society and 59 in Jalgaon society.

The Jalgaon society prices its share at the rate of Rs. 10 as against Rs. 100 in case of the Kissan society. Both the societies were having a membership of around 500 by 1992-93. However, in 1992-93, the share capital of the Jalgaon society was Rs. 17 lakh as against only Rs. 6 lakh in case of the Kissan society (Table 13.1) which was mainly due to the huge capital contribution of the Government in 1988-89. It was observed that though the Government contributes towards the share capital of the societies (repayable after 10 years), it does not interfere in their management.

It may be observed from Table 13.1 that membership registered a nominal growth of 4 per cent (Kissan) and 5 per cent (Jalgaon). Although the growth of membership was slow in the societies, the 'own' contribution has increased over the years (Table 13.1) and this may be attributed to the policy of 'ploughing back' 1 per cent of the sale proceeds towards the share capital of the members. This, in turn, strengthens the owned fund (members' share capital, reserves and other funds) of the societies. In fact, it was observed that between 1982-83 and 1992-93 the owned fund increased by 3 and 6 folds in case of Kissan and Jalgaon societies respectively.

Table 13.1 : Growth of Membership and Share Capital

Year	*KCFSSL*				*JCFVSSL*			
	Member-ship (No.)	*Share Capital (Rs. lakh)*			*Member-ship (No.)*	*Share Capital (Rs. lakh)*		
		*Own**	*Govt. loan*	*Total*		*Own**	*Govt. loan*	*Total*
1982-83	357	2.14	0.50	2.64	362	1.65	—	1.65
1983-84	375	2.53	1.00	3.53	376	1.69	—	1.69
1984-85	410	2.93	1.00	3.93	380	2.01	—	2.01
1985-86	450	3.55	1.00	4.55	475	2.31	—	2.31
1986-87	455	4.12	2.00	6.12	492	2.65	—	2.65
1987-88	480	4.21	2.00	6.21	489	3.19	—	3.19
1988-89	489	4.25	2.00	6.25	488	3.41	10.00	13.40
1989-90	500	4.27	2.00	6.27	540	4.36	10.00	14.36
1990-91	504	4.27	1.75	6.02	551	5.31	10.00	15.31
1991-92	515	4.28	1.61	5.89	549	6.37	10.00	16.37
1992-93	526	4.29	1.46	5.75	564	7.04	10.00	17.04
CGR(%)	3.81	6.68		7.43	4.74	16.71		33.91

*Includes contribution of the members and 1% of the sale proceeds plown back by the society towards the share capital.

In order to enhance the membership enrolment, we feel that the society should gradually dispense with the policy of marketing the banana supplied by the non-members (nominal members) thereby making it mandatory on the part of the grower to become the permanent member so as to sell their banana through the society. This helps the society to know in advance the quantity of banana to be supplied and accordingly its sale can be arranged. As better performance would attract more membership, the societies should strive hard to improve their performance. The suggestions that follow would probably help in this regard.

Marketing Policies

Commission Charges : The main objective of the societies is to market the banana fruits of their members on commission basis. While the Jalgaon society charges a commission of 3 per cent it is 2 per cent in case of the Kissan society. The high commission charged by the Jalgaon society was due to the policy of providing harvesting (cutting of banana) and transport facilities to their members at their field itself.

Method of Sale and its Impact : The general marketing channels followed by the Jalgaon cooperative fruit sale societies are presented in Fig. 13.1 though each society follows different methods to market banana of their members. While the Jalgaon society sells banana through wholesalers in the local market itself, the Kissan society sends most of its banana to Delhi and markets them through its commission agents.

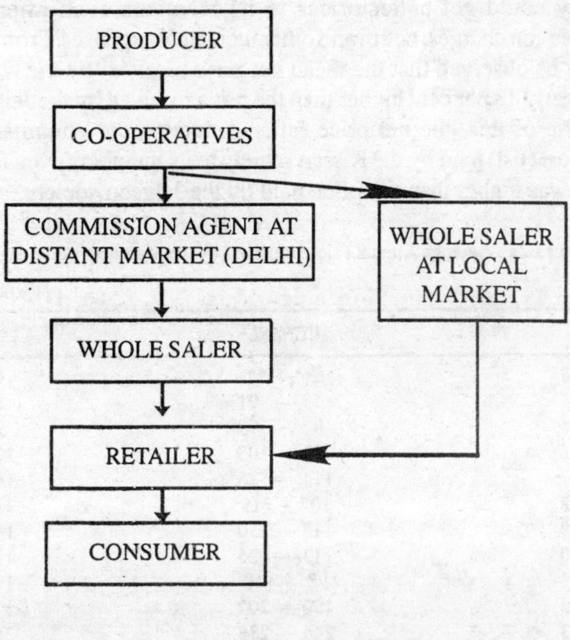

Fig. 13.1 : Marketing Channels of Jalgaon Societies

As the railway offers a concession of 40 per cent on the freight, the society usually sends banana by 'banana special' trains. However, sometimes they are forced to send by trucks due to non-availability of railway wagons which results in higher cost of transport. The produce is brought by the members to the nearby Nimbhora railway station on their own and each bunch is weighed and recorded before loading it to the wagon. After reaching the New Azadpur railway station, banana is sold on 'wagon basis' by the commission agents to the wholesalers by 'Hatha Method' (covered hand auction). The commission agents deduct their commission at the rate of 8 per cent and their expenses on account of octroi, loading/unloading and postage. However, if the society sells through them regularly for more than a year, then as a mark of good will, the commission agents return 2 per cent of the commission to the society. Thus, the net commission charged would be 6 per cent.

In order to understand as to whether the method of sale has any impact on the price, a comparison of the annual price obtained by the members of the societies for the period from 1982-83 to 1992-93 is made in Table 13.2a. Further, a disaggregate analysis of the price paid to the members and the price received by the societies in 1992-93 is attempted in Table 13.2b. The results presented in Table 13.2a and 13.2b clearly show that the Kissan society could get better prices to its members even after paying the commission charges, octroi and other incidental expenses. From Table 13.2b it may be observed that the mean net price received by the Kissan society was nearly 13 per cent higher than the price received by the Jalgaon society. Because of this, the net price (after deducting the commission and the transport cost) paid by the Kissan society to its members even for the lowest grade was higher than the price paid by the Jalgaon society.

Table 13.2a : Average Annual Prices received by the Members of the Societies (Rs/q)

(1982-83 to 1992-93)

Year	*KCFSSL*	*JCFVSSL*
1982-83	63*-75**	60
1983-84	79 — 91	82
1984-85	87 — 99	91
1985-86	93 — 105	104
1986-87	114 — 126	144
1987-88	107 — 119	111
1988-89	118 — 130	141
1989-90	113 — 125	112
1990-91	98 — 110	107
1991-92	190 — 202	178
1992-93	246 — 258	240

*Price of the lowest grade 'E'

**Price of the highest grade 'A'

Table 13.2b: Net Prices received by the Members and the Societies

(Rs/q)

Months	*Net price received by societies*		*Net price paid to members*	
	KCFSSL	*JCFVSSL*	*KCFSSL*	*JCFVSSL*
April '92	281.71	276.33	274.95* - 286.95**	267.98
May '92	205.60	200.39	200.72 - 212.72	161.28
June '92	183.51	179.89	177.85 - 189.85	174.39
July '92	169.32	269.44	158.22 - 170.22	261.24
August '92	247.60	268.07	238.69 - 250.69	259.93
September '92	238.78	212.86	217.15 - 229.15	206.37
October '92	230.65	169.66	204.15 - 216.15	164.47
November '92	228.73	177.60	199.70 - 211.70	172.52
December '92	224.51	191.20	196.51 - 208.51	185.37
January '93	217.43	159.38	163.39 - 175.39	154.50
February '93	267.86	158.47	223.16 - 235.16	153.62
March '93	230.22	150.99	187.79 - 199.79	146.36
Mean	227.16	201.19	203.52	192.33
CV (%)	13.94	22.78	15.92	23.56

*Price of the lowest grade 'E'
**Price of the highest grade 'A'

Grading and Price Policy : Both the societies follow grade based price fixation. Grading is done mostly on the basis of the weight of the bunch. In case of the Jalgaon society price is fixed for a standard weight of 10q/100 bunches and for the lot of 100 bunches weighing more than 10 quintals, an additional Rs. 10-12/q is paid above the fixed price. Similarly the lot weighing less than 10 quintals fetches Rs. 10-12/q less than the price fixed for the standard weight. The bunches weighing less than 10 kg are usually rejected by the wholesalers.

On the other hand, the Kissan society arrives at an average monthly price for the lowest grade 'E' (Grades : A = > 15kg/bunch, B = 13-15 kg/bunch, C = 11-13 kg/bunch, D = 9-11 kg/bunch, E = < 9kg/bunch) based on the net sale proceeds received from the commission agents. An increment of Rs. 3/q is made for the successive grades while making the payment to the members. As the society grades up to 'E' (which forms around 15 per cent of the total banana sold), marketing of the inferior quality banana is also taken care of which otherwise would be difficult to the members.

Supply of Inputs : Realising the importance of fertilisers in the production of banana, both the societies have been engaged in the sales of fertilisers. Fertiliser sales has been an important activity of the Kissan society and over the years the share of this activity has been in the range of 18 to 45 per cent of the total sales of the society whereas in the case of the Jalgaon society the share has been in the range of 2 to 17 per cent of the total sales though the sale has increased in 1992-93 with the opening up of the fertiliser outlets at Palaskheda, Dapora and Patri besides the Head Office in Jalgaon.

As may be seen from Table 13.3 while there has been a phenomenal growth of 18 per cent per annum in the fertiliser sales of Kissan society, the growth has been as low as 0.70 per cent in case of the Jalgaon society. In 1992-93, while the Kissan society distributed fertilisers worth Rs. 109 lakh the sale of Jalgaon society has been around Rs. 26 lakh. This is perhaps due to the less advance given by the society to its members for the purchase of fertilisers as we shall see later.

Table 13.3 : Growth in Value of Fertilisers Sold by the Societies

(*Rs. lakh*)

Year	*Value of fertilisers sold*	
	KCFSSL	*JCFVSSL*
1982-83	14.32 (28.62)	12.52 (16.63)
1983-84	14.83 (33.37)	7.52 (10.71)
1984-85	22.54 (32.52)	8.65 (9.02)
1985-86	25.38 (26.00)	8.82 (6.84)
1986-87	29.15 (31.77)	3.83 (3.57)
1987-88	27.18 (29.00)	2.08 (1.36)
1988-89	31.76 (35.38)	5.87 (3.79)
1989-90	49.47 (45.82)	2.99 (1.51)
1990-91	33.88 (18.27)	6.85 (2.56)
1991-92	63.45 (35.00)	5.95 (1.95)
1992-93	109.46 (29.71)	25.98 (9.67)
CGR (%)	18.45	0.70

Note : Figures in parentheses are percentages to total sales of the respective societies.

Linking of Credit with Marketing : In principle both the societies advance loans for the purchase of fertiliser and other inputs and link it with the sale of banana of the members. While the Kissan society provides interest free advance (which comes from the commission agents in Delhi), the Jalgaon society advances loans at the rate of 16 per cent interest. The Kissan society provides up to Rs. 5 bunch spread over 3 months viz. Rs. 2/bunch each in April and May and Re. 1/bunch in June after the supervision of the orchard and signing of an agreement by the member indicating his willingness to sell banana through the society. In case of the Jalgaon society, an amount of Rs. 1.50 to Rs. 2 per bunch is provided depending on the quality of the orchard, after supervision in April every year. While making payment, the Jalgaon society deducts the advance taken by the members besides 3 per cent commission and 20 paise/q towards weighing charge. The Kissan society deducts the commission at the rate of 2 per cent, railway freight, wagon expenditure and the advance given to the members from the sale proceeds before making the payment. Thus, as the advance amount is deducted from the sale proceeds of the members, recovery has never been a problem to either of the societies. This type of 'linking of credit with marketing' is worth emulating by such societies as HOPCOMS and FRESH, which do not at present advance loans to their members perhaps due to the fear of non-recovery [*see Subrahmanyam et al. (1979, 1994a)*].

Incentive for Participation : As an incentive for selling banana through them the societies are giving bonuses to their members. At the end of the year, the Kissan society gives a bonus of 60 paise/q and adds it to the price paid to the members and the Jalgaon society gives a rebate of 30 paise/q on the advance given to the members. Besides, the Kissan society has been declaring dividends ranging from 9 per cent to 12 per cent depending on the profit made in the year.

Financial Performance

Progress of Sales : It may be seen from Table 13.4 that over the years between 1982-83 and 1992-93, the turnover (at current prices) of both the societies has increased at a rate of around 14 per cent and 17 per cent per annum respectively in case of the Kissan and the Jalgaon societies. As may be observed from the table, even at constant prices (real terms) the turnover registered a growth of around 5 per cent and 8 per cent respectively. Of the total sales, banana sales were in the range of 54-82 per cent in case of Kissan society and 77-96 per cent for Jalgaon society. In 1992-93, banana accounted for 55 per cent of the total sales of the Kissan society and it was around 87 per cent for the Jalgaon society.

In Jalgaon district, banana is planted in two seasons viz 'Mrug Bagh' (June) and 'Kande Bagh' (October). About 70 per cent of the planting takes place in Mrug Bagh (MB). However, it was reported [*Waskar* and *Roy (1993)*] that though yields are lower in Kande Bagh, growers having irrigation (which is essential for the crop) plant banana in this season to take advantage of the higher prices due to poor arrivals.

Table 13.4 : Financial Performance of the Societies

Year	*KCFSSL*			*JCFVSSL*		
	Turnover (Rs. lakh)*	*Overheads as % of turnover*	*Net profit (Rs.)*	*Turnover* (Rs. lakh)*	*Overheads as % of turnover*	*Net profit (Rs.)*
1982-83	63.32 (77)	2.75	90742 (1.44)	75.31 (75)	3.12	108923 (1.45)
1983-84	63.46 (77)	5.62	54687 (0.87)	70.23 (78)	3.50	107095 (1.52)
1984-85	69.31 (67)	1.22	56815 (0.82)	95.86 (83)	2.93	174782 (1.83)
1985-86	97.75 (74)	1.35	69319 (0.71)	101.63 (83)	3.13	92955 (0.92)
1986-87	91.77 (68)	1.62	70548 (0.77)	107.40 (87)	5.57	73381 (0.68)
1987-88	94.23 (71)	1.70	46051 (0.50)	152.69 (91)	5.86	166268 (1.09)
1988-89	89.76 (65)	1.86	45149 (1.11)	169.04 (99)	4.73	219432 (1.30)
1989-90	107.96 (54)	1.89	93181 (0.86)	197.80 (95)	3.95	68067 (0.34)
1990-91	185.46 (82)	0.88	76169 (0.41)	267.37 (94)	3.37	292174 (1.09)
1991-92	182.10 (65)	1.89	101590 (0.56)	304.80 (96)	2.99	391792 (0.33)
1992-93	240.92 (55)	1.65	201794 (0.84)	268.76 (87)	3.82	187172 (0.70)
CGR % at current prices	13.59		7.06	16.81		9.41
CGR % at constant prices	4.86			7.82		

*Includes sales of banana and fertiliser in case of KCFSSL and banana, fertiliser and a smallquantity of vegetables in case of JCFVSSL.

Note:

Figures in parentheses are :

1. Banana sales as percentage ot turnover,
2. Net profit as percentage of turnover.

Overheads and Net Profit : The increase in the turnover of both the societies witnessed during the period from 1982-83 to 1992-93 was accompanied by an increase in the overheads also. However, by maintaining the overheads proportional to the turnover in the range of 2-3 per cent (the Kissan society) and 3-6 per cent (the Jalgaon society), societies could make profits over the years (Table 13.4). As a matter of fact, it may be seen from the table that there was not a single year when these societies incurred losses. However, it is interesting to note that the net profit as a proportion of the turnover has been 1 per cent or less in both the societies showing the true spirit of cooperative society i.e. no profit no loss principle. Further, it was observed that this profit was used to build up funds like reserve fund, building/godown fund, price fluctuation fund etc., which in turn strengthen the owned fund of the societies. Besides, the benefits are passed on to the members in the form of dividend.

Ratio Analysis : In order to understand the financial soundness of the societies certain accountancy ratios were calculated and are presented in Table 13.5. These ratios indicate that both the societies are financially sound. In fact, the owned fund formed about 85-90 per cent of the 'capital structure' (owned fund + term loans) of the Kissan society and it was 70-75 per cent in case of the Jalgaon society. However, as is evident from the quick ratios of the recent years, the Kissan society appears to be less liquid compared to the Jalgaon society. The quick ratio of the society is less than the ideal ratio of 1 : 1 and this calls for proper inventory management. It may be observed from the table that the ratio of total assets to total liabilities is just above 1 for both the societies thereby indicating that the assets are just sufficient to meet the obligations of the societies. From the ratio of current liabilities to owned fund, it is clear that both the societies have been able to meet the short term obligations with their owned fund itself. But, the ratio of total liabilities to owned fund points out that the owned fund was not sufficient to cover all the long term commitments of the respective societies thereby suggesting that there is still scope for further strengthening of their owned funds.

Conclusions

1. The Jalgaon societies' policy of marketing of perishables on behalf of the members on commission basis, instead of outright purchase as practiced by some of the societies like HOPCOMS,

Table 13.5 : Financial Ratios of the Cooperative Fruit Sale Societies

Ratio	*1992-93*	*1991-92*	*1990-91*	*1989-90*	*1988-89*	*1987-88*
	The Kissan Coop. Fruit Sale Society					
1. Current Ratio	2.4700	1.4200	2.3100	1.2800	1.3800	1.3800
2. Quick Ratio	0.7100	0.3000	1.2500	1.0400	1.1400	1.2500
3. Ratio of Total Assets to Total Liabilities	1.0700	1.0500	1.0400	1.0300	1.0200	1.0200
4. Ratio of Total Liabilities to Owned Fund	1.4900	1.1400	1.2000	2.2300	1.9700	2.1400
5. Ratio of Current Liabilities to Owned Fund	0.6400	0.2600	0.3000	1.3400	1.0400	1.2100
6. Net Profit to Turnover	0.0060	0.0060	0.0041	0.0090	0.0100	0.0050
7. Net Profit to Owncd Fund	0.0055	0.0056	0.0041	0.0086	0.0050	0.0049
	The Jalgaon Coop. F&V Sale Society					
1. Current Ratio	3.2900	2.6900	4.0600	4.3400	6.7400	6.4100
2. Quick Ratio	3.2000	2.6800	3.9600	4.2100	6.6900	6.3600
3. Ratio of Total Assets to Total Liabilities	1.0200	1.0300	1.0300	1.0100	1.0300	1.0200
4. Ratio of Total Liabilities to Owned Fund	1.6500	1.6700	1.6000	1.7200	1.5500	1.7100
5. Ratio of Current Liabilities to Owned Fund	0.2400	0.3100	0.1600	0.1500	0.0800	0.0900
6. Net Profit to Turnover	0.0070	0.0033	0.0110	0.0034	0.0130	0.0110
7. Net Profit to Owned Fund	0.0291	0.0656	0.0561	0.0148	0.0480	0.0385

Bangalore and FRESH, Hyderabad (*Subrahmanyam et al., 1994b*), appears to be a sound policy to be followed by other societies to avoid too much risk at least in the initial stages of their operation.

2. The marketing practices followed, especially by the Kissan society, have shown the helpful role the commission agents can play in marketing of fruits for exploiting the price advantage in the distant market.
3. The societies are depending only on the Delhi market and can explore the possibilities of diversifying the markets by sending to other metropolis by having tie up with other fruits and vegetables societies like HOPCOMS of Bangalore, FRESH of Hyderabad etc. In Delhi, they should establish contacts with the Mother Dairy (NDDB) so as to reduce their dependence on the commission agents.
4. The societies should try to send the graded banana fruits in separate lots to get better prices as at present banana is sold on 'wagon basis'.
5. At present banana is sold by 'Hatha Method' at the railway station itself. Instead, as per the Agriculture Produce Marketing (Regulation) Act, open auction system may be tried by taking the produce to the market yard. As a matter of fact, the Delhi Government has decided to ban the practice of 'covered hand auction' in Azadpur market to prevent corruption in marketing and to help the farmers (*Annon, 1994*).

PART V

PLANNING A SUCCESSFUL SOCIETY

14

GUIDELINES FOR PLANNING AND RUNNING SUCCESSFUL FRUIT AND VEGETABLE COOPERATIVE MARKETING SOCIETY

From the in-depth analysis of the four societies carried out in the earlier part, we could identify some of the reasons/weaknesses of both the successful and not so successful societies. Based on these findings, one can formulate some guidelines which will help in successfully running them and these are described briefly below:

Overhead Expenses

One of the main findings that emerged from the studies carried out was that the most important reason for the societies to run into losses was disproportionate overhead expenses. Societies whose overhead expenses exceeded the volume of business and trade profits have incurred losses and those societies whose overhead expenses were less than trade profits could make profits. Hence there is a need to maintain a proportionate balance between these two.

Generally, the overhead expenses come from acquiring immovable and movable assets like trucks, construction of cold storages, office buildings, godowns, cars for the managing directors, too much of staff, having a post office etc. Hence, the new societies should try to keep these expenses at the minimum. Some of the steps suggested are:

1. In the initial stages, it is better to hire the trucks for transporting the produce based on the volume of business rather than purchasing them outright from loans. This step not only saves

in blocking the capital but also saves recurring expenditure like salary of the drivers, repairs and maintenance etc.

2. The co-operative societies should not, in the initial stages, try to construct huge cold storages, as not only the construction of cold storages requires huge capital but also involves recurring expenditure for their maintenance in terms of electricity consumption, diesel consumption for generator sets etc. Hence it is desirable to hire space in the existing cold storages in the initial stages. At later stages one can take up construction of cold storages of appropriate capacity depending upon the volume and kind of commodities to be stored.
3. In the initial stages it is better to have minimum number of permanent staff and others like Accountant etc. should be on a part time basis if feasible.
4. The office accommodation and infrastructure also should be proportionate to the volume of business and should be kept at the minimum.

   In general it is desirable to avoid blocking in of capital in movable and immovable assets.
5. It is better to avoid too much capital borrowing and the principle of equity should be kept in mind. Too much borrowing will result in using the profits for servicing costs like interest etc.

Share Capital

It was observed that most of the share capital is contributed by the State Governments and the share capital received through membership is very meagre. The excess contribution by Government will make societies to be slack on the enrolment of members as evidenced from the stagnation or very slow growth in membership of the societies studied. Hence the government contribution to share capital should be proportionate to the share capital received through membership. This also avoids the feeling that it is government concern and makes the members feel that the society belongs to them.

Steps to Increase Membership

1. Though the societies may accept the produce from non-members also in the initial stages some incentive in the form of slightly higher procurement price, supply of inputs etc. should be given

to members to attract cultivators to become members which also helps to increase the share capital.

2. All the produce brought by the cultivators should be accepted and different prices based on the quality should be offered and not the extreme policy of paying uniform price or accepting only quality produce as is practiced by some of the societies studied.

Business Policies for Increasing Profits

1. The procurement of produce from the market should be kept at the minimum as far as possible and majority of the produce should be procured from the cultivators.
2. The main procurement centre at headquarters should be located at a place where most of the fruits and vegetables arrive by buses from villages or where the produce is brought for sale i.e. market places. If it is located away from these places, the cultivators will have to transport the produce twice which will double the cost of transportation.
3. It is also desirable in the short run to use the services of the commission agents at distant markets to take advantage of the higher price at these places. This step is essential till the society can start its own branch at these places.
4. The sale of members' produce on commission basis may be undertaken instead of outright purchase. But of course it is essential that the cultivators are paid some amount based on the prevailing wholesale prices.
5. Whenever co-operative societies are functioning in other places, it is desirable to establish links with them for exchange of produce and sale.
   For example produce like apples can be sent by Himachal Pradesh Co-operative Marketing Society to HOPCOMS, Bangalore, Mother Dairy, New Delhi etc., Similarly HOPCOMS, Bangalore can send some vegetables and fruits like mango to HP Society. Such types of linkages can be worked out.
6. There should not be duplication of sale counters and branches of the societies when more than one cooperative society is functioning in a district or state.
7. The linking of credit with marketing should be taken up by advancing credit in kind by supply of inputs like fertilisers, pesticides etc.

8. Inspection of members' gardens/fields for giving technical advice in collaboration with the Horticulture Department and for overseeing proper utilisation of credit (inputs) should be taken up by the societies.

Besides the above, the singlemost important factor which contributes for the success of the society is the choosing of the right type of person(s) to lead the societies who has/have inclination, will and service mind.

The above is not an exhaustive list but only main guidelines and depending upon the situation and problems, steps have to be taken to correct them.

ACRONYMS AND UNITS

ACRONYMS

ADC	:	Agricultural Development Council
AIRCRC	:	All India Rural Credit Review Committee
AIRCSC	:	All India Rural Credit Survey Committee
AP	:	Andhra Pradesh
APEDA	:	Agricultural and Processed Food Products Export Development Authority
ARC	:	Agricultural Refinance Corporation
AVRDC	:	Asian Vegetable Research and Development Centre
BGGCOMS	:	The Bangalore Grape Growers' Cooperative Marketing and Processing Society Ltd.
BHOPCOMS	:	The Bangalore Horticultural Producers' Co-operative Marketing and Processing Society Ltd.
CARDS:	:	Centre for Agricultural and Rural Development Studies
CGR	:	Compound Growth Rate
CIS	:	Commonwealth of Independent States
CL	:	Current Liabilities
CLOFR	:	Current Liabilities to Owned Fund Ratio
CR	:	Current Ratio
CV	:	Coefficient of Variation
DKHOPGCOMS	:	The Dakshina Kannada District Horticultural Produce Growers' Cooperative Marketing and Processing Society Ltd.
EEC	:	European Economic Community
FA	:	Fixed Assets
FAO	:	Food and Agriculture Organisation
FAOFR	:	Fixed Assets to Owned Fund Ratio
FATAR	:	Fixed Assets to Total Assets Ratio
FIG	:	Figure
FRESH	:	Framers' Rural Extension Service in Horticulture

		Cooperative Marketing Society, Ltd.
GOI	:	Government of India
GOK	:	Government of Karnataka
HOPCOMS	:	The Horticultural Producers' Cooperative Marketing and Processing Society, Ltd.
HP	:	Himachal Pradesh
HPMC	:	Himachal Pradesh Horticultural Produce Marketing Corporation
HRD	:	Human Resource Development
IATA	:	International Air Transport Authority
ICAR	:	Indian Council of Agricultural Research
ICMR	:	Indian Council of Medical Research
IDA	:	International Development Agency
IIHR	:	Indian Institute of Horticultural Research
IIM	:	Indian Institute of Management
IIMC	:	Indian Institute of Management, Calcutta
IIP	:	Indian Institute of Packaging
ISO	:	International Standards Organisation
JCFVSSL	:	The Jalgaon Cooperative Fruits and Vegetables Sales Society Ltd.
KCFFSSL	:	The Kissan Cooperative Fruits Sales Society
MB	:	Mrug Bagh
KB	:	Kande Bagh
KNK	:	Karnataka
MIS	:	Market Intervention Scheme
NABARD	:	National Bank for Agriculture and Rural Development
NAFED	:	National Agricultural Cooperative Marketing Federation of India
NDC	:	National Development Council
NCA	:	National Commission on Agriculture
NCDC	:	National Cooperative Development Corporation
NCMS	:	Nilgiris Cooperative Marketing Society
NDDB	:	National Dairy Development Board
NGO	:	Non-Government Organisation
NHB	:	National Horticulture Board
NP	:	Net Profit
NPTOR	:	Net Profit to Turnover Ratio
NVGC	:	Nilgiris Vegetables Growers' Cooperative Marketing Society

PAU	:	Punjab Agriculture University
PHC	:	Pre Harvest Contractor
PPC	:	Plant Protection Chemicals
PSS	:	Price Support Scheme
QA	:	Quick Assets
QR	:	Quick Ratio
RBI	:	Reserve Bank of India
RCA	:	Royal Commission on Agriculture
SFDA	:	Small Farmers' Development Authority
TA	:	Total Assets
TATLR	:	Total Assets to Total Liabilities Ratio
TLOFR	:	Total Liabilities to Owned Fund Ratio
TN	:	Tamil Nadu
TNAU	:	Tamil Nadu Agriculture University
TPD	:	Tonnes per Day
UP	:	Uttar Pradesh
UAS	:	University of Agricultural Sciences
USA	:	United States of America
UT	:	Union Territory
VEFCO	:	Vegetable and Fruit Cooperative Marketing Society Ltd.

UNITS

ha	:	Hectare
kg	:	Kilogram
km	:	kilometre
ml	:	Millilitre
q	:	Quintal (100 kg = 1q; 10 = 1t)
Re	:	Rupee
Rs	:	Rupees
t	:	Tonnes

BIBLIOGRAPHY

Abbot, J.C., 1970. *Marketing Fruits and Vegetables*, FAO Marketing Guide No. 2, FAO, Rome.

Ananth, G.S., 1984. *Performance Appraisal of the Bangalore Grape Growers' Cooperative Marketing and Processing Society Ltd., Lal Bagh, Bangalore,* M.Sc. (Ag) Thesis, UAS, Bangalore.

Anon, 1992. Vital Statistics, *Agricultural Marketing*, Oct-Dec.

Anon, 1994. Azadpur Covered Hand Auction, *Economic Times*, 26-3-1994.

APEDA, 1995. *Financial Assistance Schemes*, APEDA, New Delhi.

Asha, 1989. A Study of Price and Supply Variability of Vegetables in Varanasi Market, *Agricultural Marketing*, 32(2) : 16-17.

Attwood, D.W. and B.S. Baviskar, 1988. *Who Shares? Cooperatives and Rural Development*, Oxford University Press, New Delhi.

Ayyasamy, P.K., 1980. Distress Sales in Tomato—Tomato producer gets only 50% of what the consumers pay, *Indian Express*, January 24, p. 9.

Bale Rao, M.M., S.L. Ansari and B.P. Tyagi, 1981. Marketing of Vegetables—A Sample Study, *Agricultural Marketing*, 24(1) : 27-29.

Bhatnagar, A.P., 1985. Growth of Cold Storage Industry in India, *Agriculture Marketing*, 28(8) : 69-72.

Biradar, K.S. and D.V. Kasar, 1983. Marketing of Jalgaon Banana in Delhi Market, *Indian Journal of Marketing*, 13 (6 & 7) : 27-30.

Biradar, K. S. and D.V. Kasar, 1984. A Study of Relative Efficiency of Cooperatives vis-a-vis other Agencies in Marketing of Jalgaon Banana in Delhi Market, *Indian Cooperative Review*, 22(2) : 231-36.

Calkins, P.H. and Hu-mei Wang, 1978. *Improving Marketing of Perishable Commodities—A Case Study of Selected Vegetables in Taiwan*, Technical Bulletin No. 9, AVRDC, Taiwan.

Chadha, K.L., 1994. Horticulture : Delayed Recognition of Potential, *Hindu Survey of Indian Agriculture*, pp. 111-17.

Chadha, K.L., 1995. India's Fruit and Vegetable Scenario, *Horticultural Information Service*, 1(11) : 108-17.

Chand, K.P., R. Singh and M.L. Sharma, 1986. Determining the Success of Cooperative : A Study in H.P., *Indian Cooperative Review*, 24(1) : 46-55.

Das, A.K. and Sugata Ghosh, 1996. Marketing of Mango in Vijayawada, *Indian Horticulture*, 41(2) : 12-17, FAO, 1994, Production Year Book.

Gajanana, T.M. and K.V. Subrahmanyam, 1993. Cooperative Marketing of Fruits and Vegetables—Present Status and Future Requirements, *Indian Cooperative Review*, 31(1) : 34-38.

Gajanana, T.M. and K.V. Subrahmanyam, 1996. Fruit Marketing Through Cooperatives : A Case Study of Jalgaon Banana, *Agricultural Economics Research Review*, 9(1) : 36-48.

GOI, Draft Five Year Plans (Various volumes).

Gopal, K.N., 1978. *An Analysis of Problems of Marketing of a few Vegetables in Bangalore City*, M.Sc. (Ag Econ) Thesis, UAS, Bangalore.

Government of Karnataka, 1992. Annual Report, Department of Corporation, Bangalore, Karnataka.

Gupta, V.K. and Ramesh Gupta, 1985. *Kanchan and its Strategy for Growth*, FAO and IIM, Ahmedabad.

Hajella, T.N., 1978. *Principles, Problems and Practice of Cooperation*, Shivlal Agarwala & Co., Educational Publishers, Agra.

Hugar, L.B., S. Suryaprakash, and D.S. Lokamanya, 1983. A Study of Costs and Margins in Marketing of Brinjal in Belgaum City, *Agricultural Marketing*, 26(1) : 23-30.

IIHR, 1992. Estimation of Post-harvest Losses of some of Horticultural Crops and their Economic Implications, *IIHR News*, 13(1) : 1-3.

Kahlon, D.S., 1988. Role of Marketing Cooperatives with special reference to Integrating Credit with Marketing, *Indian Journal of Agricultural Marketing*, 1(1).

Kaul, G.L., 1993. Development of Horticulture : A Boost in the VIII Plan, *Indian Horticulture*, 38(1) : 9-13.

Krishna, P.V., 1976. Marketing of Fruits through Cooperative Society—A Case, *Indian Journal of Agricultural Marketing*, 7(1) : 27-32.

Kuchhal, S.C., 1988. *Financial Management : Analytical and Conceptual Approach*, Chaitanya Publishing House, Allahabad.

Mamoria C.B. and R.D. Saksena, 1973. *Cooperation in India*, Kitab Mahal, Allahabad.

Mehta, S.W., 1989. Role of HPMC in the Introduction of Plastic Crates in Marketing of Apples and Oranges, *Indian Journal of Marketing*, 19(6 & 7) : 7-8.

Mehta and B. Singh, 1973. *Marketing of Citrus Fruits in Punjab*, Department of Economics and Sociology, PAU, Ludhiana, (Mimeo)

Mishra, R.B., 1972. Cooperative Cold Storage—Some Lessons from Bihar, *Indian Cooperative Review*, 10(1) : 73-84.

Mishra, R.B., 1975. The Development of Cooperative Agricultural Marketing—Some Real Constraints, *Indian Cooperative Review*, 13(1) : 25-38.

Mohandoss, V., Mruthyunjaya and K.V. Subrahmanyam, 1979. Study of Fruit and Vegetable Cold Store Units—A Comparative Performance of Different Agencies, *Indian Journal of Agricultural Economics*, 34(4) : 179-185.

Mohandoss V., Mruthyunjaya and K.V. Subrahmanyam, 1981a. Capacity

Utilization Concepts and their Relevance to Costs and Returns of Fruit and Vegetable Cold Stores, *Indian Journal of Agricultural Economics*, 36(2): 56-59.

Mohandoss, V., Mruthyunjaya and K.V. Subrahmanyam, 1981b. A Study of Fruit and Vegetable Cold Storage Units in Bangalore City, *Agricultural Situation in India*, 35(10) : 765-772.

NABARD, Statistical Statements Relating to Cooperative Movement in India (Non-Credit Societies).

NAFED, Annual Reports.

NCDC, Annual Reports.

NHB, 1995, *Horticulture Data-Base,* NHB, New Delhi.

NHB, 1995, *National Horticulture Board—A Profile*, NHB, New Delhi.

Narasimhan, M.K. and N.S. Ravindra, 1989. Role of Cooperative Society in the Marketing of Vegetables in Bangalore City, *Indian Journal of Agricultural Marketing*, 3(2) : 126.

Nawadkar, D.S., D.L. Sale and J.G.R. Pawar, 1990. Marketing of Fruits through Producers' Associations vis-a-vis Traditional Traders—A Case Study in Maharashtra, *Indian Journal of Agricultural Marketing*, 4(1) : 91-92.

Parmar, G.R., G.D. Naik and K.S. Patil, 1989. Selling of Chiku and Price Received by Chiku Growers from Different Purchasing Agencies in South Gujarat, *Indian Cooperative Review*, 27(2) : 202-209.

Patil, H.N., P. Kumar and M.A. Muralidharan, 1983. Marketing Margins and Price Spread in the Marketing of Alphonso Mango in Ratnagiri District (Maharashtra), *Indian Journal of Marketing*, 14(4) : 21-24.

Patil, H.N., P. Kumar and M.A. Muralidharan, 1985. Marketing Efficiency of Cooperative Organisations, *Indian Cooperative Review*, 22(4) : 419-24.

Pawar, P., 1991. Marketing of Banana : Factors affecting Prices and Impact of Cooperative Marketing—A Case Study in Jalgaon District of Maharashtra, *Marketology*, 23 (2, 3) : 19-26.

Pawar, P., S.N. Tilekar and B.J. Hinge, 1989. Sale of Jalgaon Banana through Organised and Unorganised Marketing Sectors, *Indian Journal of Agricultural Marketing*, 3(12) : 23-24.

Pradhan, M.S., 1968. Cold Storage in Cooperative Sector, *Indian Cooperative Review*, 5(4) : 537-542.

Pradhan, M.S., 1989. Cooperatives in Vegetable Marketing—Present Status and Future Outlook, In : *Proceedings of the National Symposium on Post-harvest Management of Vegetables*, NHB, New Delhi, pp. 127-30.

Puri, S.S., 1992. Cooperative Marketing, *Indian Cooperative Review*, 30(2) : 101-110.

Raghunandan, A., 1992. Diversifying Agriculture by boosting Horticulture, *Financing Agriculture*, 24(4) : 4-8.

Rajagopal, 1996. Strengthening Agribusiness Cooperatives : NGO Intervention Model, *Southern Economist*, 35(4) : 11-12.

Ramasamy, C. 1980. Problems of Production and Marketing of Major Vegetables in Coimbatore District (1) Brinjal—Research Report, AgEcon/2/18, CARDS, TNAU, Coimbatore.

Ranade, C.G., D.P. Mathur, B. Rangarajan and V.K. Gupta, 1988. *Performance of Integrated Milk Cooperatives*, Concept Publishing Company, New Delhi.

Ranganna, B. *et al.*, 1997. Packaging Technology for Fruits, *The Kissan World*, 24(2) : 35-37.

Reddy, B.R., 1988. Financial Management in Cooperative, *The Cooperator*, 26(7) : 157-61.

Roy, S.K. and R.K. Pal, 1993. Use of Plastics in Post-harvest Technology of Fruits and Vegetables—A Review, *Indian Food Packer*, 47(1) : 27-45.

Selvaraj, K.N. and S. Krishnamoorthy, 1990. Marketing Operations for improving Marketing Efficiency of Vegetables, *Indian Journal of Agricultural Marketing*, 4(2) : 205-8.

Sharma, Keshav, 1991. *Marketing Management of Horticultural Produce*, Deep and Deep Publications, New Delhi.

Singh, B. and D.S. Sidhu, 1976. *Marketing of Mangoes in Punjab State*, Department of Economics and Sociology, PAU, Ludhiana (Mimeo).

Singh, H., 1983. Horticulture Industry and Substitution of Packaging Cases, In : *Proceedings of National Workshop on Substituting forest material timbers for Packaging of Horticulture Produce, Tea, Tobacco and Textiles*, New Delhi, 30-31st July, 1983.

Singh, H.P., 1995. Recent Developments in Direct Marketing of Fruits and Vegetables in India, *Horticulture Information Service*, 1(3) : 95-98.

Singh, J.P., 1996. NCDC for Development of Cooperatives, *The Cooperator*, 34(5) : 181-86.

Southworth, H (ed), 1974. *Some Studies of Fresh Fruit and Vegetable Marketing in Asia*. A/D/C, New York, USA.

Spinks, G.R., 1972. Myths about Agricultural Marketing, In : ADC Training Program, No. 15, New York, USA.

Subrahmanyam, K.V. 1981a. Effect of Marketing Costs on Cost of Production and Returns : Case Study of Vegetables, *Agricultural Marketing*, 24(3) : 7-9.

———; (1981b) Marketing Fruits : Self Marketing *vs.* Contracting, *Indian Journal of Marketing*, 11 (7) : 13-17.

———; 1982a. Efficiency of Different Channels in Marketing of Cabbage, Carrot and Cauliflower—A Case Study for Madurai Dist. Tamil Nadu, *Agricultural Marketing*, 25(3) : 1-5.

———; 1982b. Economics of Production and Marketing of Important Vegetables in Karnataka *The Lal Baugh*, 27(2) : 7-16.

———; 1983. *Economics of Production and Marketing of Important Fruits and Vegetables in Karnataka Andhra Pradesh and Tamil Nadu*, Division of Economics and Statistics, Indian Institute of Horticultural Research, Hessaraghatta, Bangalore. (mimeo).

Subrahmanyam, K.V. 1985, Mango Exports: Problems and Prospects, *Agricultural Situation in India*, 40(6) : 507-14.

———; 1986a. Post-harvest Losses in Horticultural Crops : An Appraisal, *Agricultural Situation in India*, 41(5) : 339-343.

———; 1986b. Marketing of Fruits : Self Marketing *Vs.* Contract Sale—Case Study of Mango Marketing in Madurai District (Tamil Nadu), *Agricultural Marketing*, 29(3) : 21-24.

———; 1987. Economics of Production and Marketing of Important Vegetables in Karnataka, *The Lal Baugh*, 29(1) : 43-52.

———; 1988. Marketing of Horticultural Crops in Karnataka—Present Status and Future Requirements, *Indian J. Agri. Marketing*, 2(1) : 15-22.

———; 1988. "Marketing of Vegetables : An Interstate Comparison of Practices and Associated Costs", *The Indian Journal of Horticulture*, 45 (1 & 2) : 153-159.

———; 1989. *Economics of Cultivation of Horticultural Crops in South India*, Technical Bulletin No. 7, IIHR, Bangalore

Subrahmanyam, K.V., Mruthyunjaya and V. Mohandoss, 1979. *Institutional Arrangements for Marketing of Fruits and Vegetables—Case Study of The Bangalore Grape Growers' Cooperative Marketing and Processing Society Ltd.* Division of Economics and Statistics, IIHR, Bangalore.

Subrahmanyam, K.V. and M. Sudha, 1993. Economic Feasibility of Establishing a Small Scale Cooperative Processing Unit in Rural Area : Case Study of Tomato, *Indian Journal of Agricultural Marketing*. 7(1) : 23-31.

Subrahmanyam, K.V., T.M. Gajanana and M. Sudha, 1994a. Cooperative Marketing of Fruits and Vegetables : Policies and Performance—A Case Study of HOPCOMS, Bangalore, *Indian Journal of Agricultural Marketing*, 8(2) : 186-93.

Subrahmanyam, K.V., M. Sudha and T.M. Gajanana, 1994b. Planning a Successful Fruit and Vegetable Cooperative Marketing Society : Lessons from the Existing Societies, *Agricultural Marketing*,, 37(2) : 36-39.

Sudha, M. and K.V. Subrahmanyam, 1996. Marketing of Fruits and Vegetables through Cooperatives—Case Study of FRESH, Hyderabad, *Indian Cooperative Review*, 33(3) : 255-67.

Swaroop, R. and Ranveer Singh, 1987. Impact of Regulation on Fruit Marketing in Northern India, *Indian Journal of Agricultural Marketing*, 1(1) : 31-36.

Tilekar, S.N. and P. Pawar, 1989. Imperfections and Price Differential in Banana Marketing, *Indian Journal of Agricultural Marketing*, 3(1) : 19-24.

Tilekar, S.N. and K.V. Bilonikar, 1993. Marketing Efficiencies of Vegetable Cooperative Marketing Societies in Ahmednagar District (MS), *Indian Cooperative Review*, 30(4) : 402-06.

Tilekar, S.N., B.J. Hinge and K.V. Bilonikar, 1993. Transport Efficiency in Marketing of Fruits and Vegetables through Cooperatives, *Indian Journal of Agricultural Marketing*, Conference Special, pp. 37-38.

Thorat, V.A., H.N. Patil and S.G. Borude, 1990. Efficiency of Self Marketing *vs.* Contracting for Alphonso Mangoes in Maharashtra, *Agricultural Marketing*, 33(2) : 46-48.

Uppal, D.K. 1994. Status Paper on Horticulture and Role of NHB in Creation of Infrastructure, In : Proceedings of the International Seminar on "Opportunities for Collaboration in Fruit and Vegetable Processing Industry in India", APEDA, MFPI, NHB, CIFTI, IIMC and EEC, New Delhi, 2-3, December, 1994.

Wadkar, S.S., J.M. Talathi and G.G. Thakare, 1994. Price Spread and Market Channels of Alphonso Mango in Ratnagiri and Sinddurg districts of Konkan Region (MS), *Indian Journal of Agricultural Marketing*, 8(2) : 250-57.

Waskar, D.P. and S.K. Roy, 1993. Studies on Harvesting, Handling, Transportation and Marketing of Banana, *Indian Food Packer*, 47(4) : 29-38.

INDEX